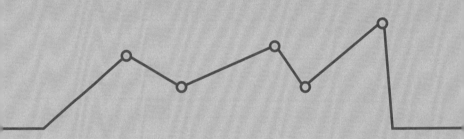

Breakthrough

Breakthrough

How to Think Like a Scientist, Learn How to

Fail and Embrace the Unknown

DR CAMILLA PANG

VIKING

an imprint of

PENGUIN BOOKS

VIKING

UK | USA | Canada | Ireland | Australia
India | New Zealand | South Africa

Viking is part of the Penguin Random House group of companies
whose addresses can be found at global.penguinrandomhouse.com

Penguin
Random House
UK

First published 2024
001

Copyright © Dr Camilla Pang, 2024

Set in 12/14.75pt Bembo Book MT Pro
Typeset by Jouve (UK), Milton Keynes
Printed and bound in Great Britain by Clays Ltd, Elcograf S.p.A.

The authorized representative in the EEA is Penguin Random House Ireland,
Morrison Chambers, 32 Nassau Street, Dublin D02 YH68

A CIP catalogue record for this book is available from the British Library

ISBN: 978-0-241-54533-1

www.greenpenguin.co.uk

I dedicate this book to my Chinese grandparents,
who I loved beyond words:

My late grandfather, Pang Cheung Fook (d. October 2023), and
my late grandmother, Pang Siu Ying (d. July 2023)

Contents

Introduction

I wasn't scared about turning thirty until the day it happened, and my father looked at me with a grin on his face. 'You don't need training any more,' he said. This was it: I'd graduated, arriving at the moment where you blink three times and become a fully functioning adult, with the last pair of stabilizers removed from your bike. As a family we'd survived the always interesting experience of my growing up with Autistic Spectrum Disorder (ASD) in a world that often struggles to accommodate people who diverge from the norm. I'd made it through school (somehow), got the PhD, written a book, and even left the family home behind and started making my own. I was, unavoidably, grown up.

None of this worried or intimidated me, except the comment that kept burrowing away in my head for days after the last of the cake was eaten. No more training. Because while I liked the idea of being a proper grown-up, which is what my dad had meant, I was scared by the thought that this might mean ceasing to discover new things. After a life spent hoovering up knowledge and trying to figure out every pattern and system that I could see and calling it living, was this it? Would my appetite for learning dwindle along with my metabolism? Would my ability to stay on the move mentally diminish as dramatically as my 5k times?

The slight panic that ensued brought me back to the abiding source of comfort in my life – back to science. As I wrote in my memoir *Explaining Humans*, science was my saviour during the formative years of my life. In a world that made little or (often) no sense to my autistic brain, I found my way using science. Like

I

the cold side of a pillow, I sought sweet relief in the certainties of science's theories and laws when human behaviour seemed completely fuzzy, defined by a relentless ambiguity that filled my head with static and my cheeks with hot frustration.

That remains the case today, but as I have turned my childhood love of the subject into a professional career, my relationship with science has also evolved. Where I once comforted myself with solid proofs and laws, I increasingly recognized that things are not so clear-cut: there are also anomalies, exceptions and outlier examples that the rules of science can't explain. Theories evolve and eventually get improved upon or displaced. New ideas cast doubt upon what we thought we knew, as well as opening up fresh angles of exploration (and sometimes, as we will see, new worlds and even universes).

As I completed my PhD researching protein structures and started working in machine learning (mostly while listening to Ella Fitzgerald), I found that science isn't so much a boiled sweet to suck on, getting you closer and closer to the gooey centre, as candyfloss that breaks apart between your teeth – equally sweet but a lot messier and more complicated, leaving you fiddling around with toothpicks for the rest of the day.

I went from thinking of science as an armour I could wrap around myself to understanding that it is first and foremost a toolbox – a combination of knowledge, techniques, disciplines and attitudes that equip you to explore the world, rather than a fixed set of points on the map. Instead of lessening my love for science, this has simply increased it. Accepting the uncertainties inherent in science has only made me hungrier to discover more and cast my net wider. I have laboured over cell cultures, trying to produce proteins from the tiniest amounts of clear liquid, spent weeks studying cancer cells to see what the engineering of a single gene would achieve, and built computer models to explore patterns in the evolution of cancers and the spread of viruses. That is the joy of science in my life: it allows me both to explore a massive canvas and to scrutinize a tiny detail, and to connect the two: figure out how the individual protein behaves and you might better understand the overall movement of the cell, and be able to target a treatment more precisely. Everything is connected, which makes the smallest nuances potentially critical to solving the biggest problems.

In the process, I've become more convinced than ever that science isn't just the best method we have for understanding the world. It's also one of the best ways we can learn about ourselves. A scientist's basic purpose is to find answers – improving how we understand every corner of our universe and the things in it. Our own lives aren't so different: we also spend most of them looking for answers in one form or another. We want to find people – and often one particular person – to spend that life with. We want to discover a career that will fulfil us as well as our bank account. We want interests, passions, experiences, good habits and bad nachos. We want the best version of our life, and that doesn't come easily for anyone. It means lots of trial and error, tears and recriminations on the way to finding things and people that just fit, even if they sometimes annoy the hell out of us.

But here's the rub. If we're being honest with ourselves – and I say this with love, in the kindest possible way – we're generally not very good at any of this, myself very much included. Too often it's all error and no trial: mistakes that get repeated, old habits we struggle to break, and an overwhelming desire to procrastinate whenever the time comes to actually try something new or difficult. This isn't necessarily our fault. No one sits a GCSE or does a degree on how to live a good and productive life. The knowledge isn't passed down like a trusty recipe for chicken soup. Instead we're left to work it out, trying to apply common sense plus the good and bad lessons we've picked up from family, friends and people we read about in magazines/ Instagram/TikTok (delete as appropriate for your generation). We muddle through and make do.

It's funny that when it comes to the most important bit of our lives – who we want to be, what we want to do, how we want to spend our time on earth – we have the tendency to DIY it. If your car breaks down you call a mechanic, and if your blood pressure needs taking then you ask a medical professional. But when it comes to figuring out your life and finding the answers it may take a lifetime to seek, we just decide to go ahead and trust our gut.

The good news is that we don't have to accept this accidental approach to life's big questions. We can in fact enlist an expert adviser: the collective knowledge of the scientific profession. Now before you scratch your head and wonder about taking life advice from people who can't necessarily be guaranteed to leave the house wearing two of the same shoes, let me be clear. My case isn't that scientists have got everything sorted, and we simply need to copy their homework. It's that they do the things we are all trying to do – looking for answers, sorting through options, making decisions and generally figuring things out – for a living. Scientists are professional question askers and truth seekers. They know how to plan an experiment, see it through and produce an honest assessment of the results. They also know how to spot problems, bring in expertise they don't have themselves, and overcome the inevitable setbacks along the way. They are familiar with troubleshooting errors, used to avoiding traps like bias, and your friend when it comes to spotting the difference between trends and anomalies (like knowing whether you actually like a person or just had a good time with them on the second date).

Scientists don't have all the answers about how to live your best life, but there's no one better when it comes to asking the right questions, setting up the experiment, and making sense of the evidence it produces. So to write this book – one that hopes to show how an understanding of the scientific mind and process can help us all answer life's big questions – I canvassed their opinions. I've taken the guidance of over a dozen brilliant scientists who talked to me about their work, how they do it, what they have learned in managing uncertainty, and what motivates them. As well as being profound, their expertise is wide-ranging: from cancer biology to data science, theoretical physics to engineering, psychology, nanoscience, machine learning and the philosophy of science. I have used many of their stories to illustrate different sections of the book, as well as their wisdom on how they think about their work, and what being a scientist has taught them. I've asked the questions I always wanted to, going from the cutting edge of research to the age-old work that provided its foundations. Here, I also draw on the history of science and some of its most notable breakthroughs – from the story of how the atom can be split to the development of quantum theory, Einstein's theory of special relativity and the search for the missing bits in the universe which we call dark matter. All these examples help to show what is involved in the pursuit of scientific discovery, and what we as humans can learn from it.

The people whose stories I will tell in this book have had brainwaves so dramatic that they ran back to their office late at night to continue working, changed the course of expensive research programmes halfway through, toiled away for years in subject areas that others deemed a waste of time, and had big ideas when building Lego models with their children or after a snack bar melted in their pocket. They show that science can happen anywhere and at any time – that breakthroughs can take an afternoon to achieve or an entire career to build towards.

And those are just the tales that have a happy ending. The story of science is as much one of failure as success. In fact, it's primarily a story about things that didn't work, or sort of worked, or yielded just a fragment of a conclusion that could be built upon. The decades-long pursuit of a big idea can turn out to be the search for something that never existed. Sometimes acclaimed experts in their field get obsessed with ideas that don't work or concepts that cannot be proven. Even Einstein was known to be wrong (well, some of the time).

The stereotype of science is that it's all order and process, theories and laws, a series of steps from which we cannot deviate. I have often relished this sense of certainty. However, the truth is so much messier and more interesting. Of course there is a huge amount of rigour, knowledge and technique involved (after all, you can't break the rules until you've learned them properly). But being a scientist is fundamentally a creative act. It's about being able to poke your finger in the world and see what you can make or break. To observe something and decide that you want to know how it works, or how its function can be replicated. To determine whether we can borrow a feature of one system and introduce it to another, or what we can learn simply from the obsessive study of how an electron moves, a cancer tumour grows or a galaxy shifts through time.

One of my hopes for this book is that it will give more people an insight into how fascinating a life in science can be: all the different forms it can take, and all the amazing work that scientists are doing – whether to probe the secrets of intelligence, develop the body's defences against life-threatening disease, explore the mysteries of the universe (possibly many universes, as we'll see), or work out what subatomic particles are up to while no one is looking. To pay a great and deserving tribute to the essential minds on whose discoveries we have built our modern way of life.

Equally importantly I want to demystify. To show what is involved in the scientific process, some of the key steps that scientists go through, and how these allow them to turn observations into hypotheses, hypotheses into experiments, and experiments into outcomes that can be used in the world. This is something we should all understand, not just out of curiosity but because the journey of scientific ideas has plenty to teach us. The thesis of this book is that we could all benefit from being a bit more scientific. Not by swapping our WFH loungewear for a badly fitting lab coat, or by consciously flexing buzzwords like 'rationality' to win an argument. But by getting in tune with the way that scientists think, the techniques that they use, and the innate curiosity that drives them. I want to recommend a scientific approach to attacking life's challenges: not being easily deterred by failure, keeping an open mind when we are criticized, being aware of our biases and not falling in love too much with our own ideas.

If the suppressed trauma of a double science lesson on a Friday afternoon is quickly rising to the surface, then don't worry. This book will not feature any equations (OK, well, one but it is small, I promise!), nor require you to invest in a Bunsen burner or to recite the periodic table. It's not another science lesson, and there is no test at the end. It's not an end-to-end exploration of the scientific process, but hopefully a journey you can enjoy and make your own. We will start at the beginning with observation and hypothesis, but respecting that the research process is often more circular than linear, we will also find ourselves tracking back and having to start again (not so different from the rest of life, then). It can get messy. Welcome to the dark side.

We will be taking a guided tour, at pretty high speed, through a scientist's way of working: the processes they go through, the attributes they need and the behaviours they often show. We will look at how scientists observe, probing for things that don't make sense or which have potential flaws in the consensus view. We'll explore how those observations are organized into a hypothesis that can be tested, how researchers choose to focus that inquiry, the way they interpret the evidence that emerges, and the thorny question of how (and whether) an idea can ever be proven. In taking that journey to the centre of the scientific process, we'll look at both enablers of and barriers to scientific nirvana: the fundamental importance of collaboration and teamwork in what is often wrongly regarded as an individual pursuit; the inevitability of encountering error and failure (and what to do about it); and the double-edged sword that is human bias in all its many forms. Finally we will take a flight of fancy together into the furthest reaches of scientific exploration, to the limits of our understanding about the universe, where the practitioners of quantum theory and its offshoots are changing how we think about pretty much everything.

In the process, we'll see that science carries important lessons for us all, which don't require us to develop a deep grasp of complex theorems. It is the practice of science that we will focus on above all, as a method of inquisition that encourages our curiosity, helps us to see the world in a different way, and both affords flexibility and imposes discipline on how we think. This isn't science as you learned it at school. It's about how scientists pursue the tightrope walk that leads to important work: immersing themselves in the consensus view while also looking for chinks in its armour; showing faith in their own ideas while also being open to feedback and criticism; setting a course towards gathering evidence and hunting down proof, but also being willing to change direction based on what they find.

The scientists whose stories I will tell in these pages have lived in different eras, studied different subjects and taken contrasting approaches. Some have swung at massive, world-encompassing ideas while others have spent their careers perfecting their understanding of one detailed niche. They cover the span of science's broad church, but also have something in common. They all have much to teach us about how to think for ourselves, spot opportunities, take calculated risks, approach new ideas and be accountable for our successes and failures. Those aren't just the ingredients of scientific good practice, but of a robust and fulfilling approach to life in general. In the course of this book, I hope to show you that a little science can spice up your life – it won't only make you feel and sound a bit smarter, but will give you the ability to approach the big questions and decisions in your life with more confidence and clarity. Because you don't need a PhD to think like a scientist, or a pair of lab goggles to change the way you see and understand the world. Nor do you need to be good with numbers. A willingness to learn and the curiosity to find out are the only credentials or equipment that this journey will require. Being a scientist isn't an easy gig, but I can assure you it is a wonderful one, with much to teach us not just about the research process, but life itself.

1. Observation

How to see the world scientifically

Ever since I can remember, I have wanted to know how things worked. I would (and still do!) take things apart – either with my hands or in my head – moving the pieces around and trying to understand how they fit together. I would take the wheels off toy cars and spend hours staring at bugs in our garden. Like a lot of people who grow up to be scientists, I had some urge to understand the world around me – not simply to take it for granted but to pull at the strings, to unravel puzzles, and to play around with the mess that I had created.

Sometimes children like this are described as inquisitive or curious. At other times adults roll their eyes and say they have too many questions, or remind them what curiosity did to the cat. But if you are lucky enough to do your PhD and become a scientist for real, those objections start to fall away. As I have found, it becomes normal to come into the lab on a Sunday to look after your cell cultures, and to dream about how a tiny volume of liquid in a plastic well (one of ninety-six on your tray) could produce a beautiful protein structure that might hold the key to tweaking a new drug or honing a vaccine. You are allowed to read a dozen different papers, all of which argue they have the 'best' method for conducting a particular experiment. That's not an esoteric way to spend time – it's your job. As a professional scientist, your curiosity no longer gets seen as noteworthy and potentially problematic. You can spend your day staring at stuff without anyone batting an eyelid. We even have a special adult word for it: observation.

Observation is the starting point both for doing science and for understanding how it is done. It's how we engage the senses, question every side of what we experience, and start to see new possibilities. It's where every single scientist starts out, whether you are taking in the world and creating your place within it, or pursuing scientific inquiries within fancy labs fuelled by grant funding and Costa loyalty cards. The root of every major discovery in science has been an observation: about something that might work, something that doesn't work, a piece that's mysteriously missing or a possibility that hasn't yet been considered. These observations are sparks, lighting the fire of an idea to pursue, a hypothesis to test, a theory to develop and a conclusion to hone. They are the beginning of the entire worlds that scientists have built from scratch.

Observation is also a subject close to my heart: being neurodivergent, you are constantly both the master of your senses and at the mercy of them. Because you experience life in vivid and acute detail, you will often see things that others miss. However much you try to mask your differences or to think 'normally' (which doesn't work, by the way), it won't stop you from picking up on things that initially make no sense. Having a brain like mine means spending every day in a perpetual spin of absorbing data and detail, and trying to make sense of what you have seen. Or, as I like to call it, being alive.

Observation sounds like a simple enough thing: you look at something, scrutinize it, pick holes and derive insights. Yet it is also so much more than a scientific starting line. It is through observing that we learn, develop knowledge, and in turn gain the ability to see more clearly – observing more acutely than we did before. The circular relationship between observation and knowledge, the way they feed each other like a wave gathering speed, is the basis of any scientific career, and indeed the human condition that pushes us to make sense of the world. As the legendary physicist and Nobel Prize winner Richard Feynman put it: 'What I cannot create, I do not understand.' This is perhaps the most famous saying of one of the twentieth century's most outstanding scientists. When he died in 1988, the words were still written on the blackboard in his classroom at Caltech, directly above another motto: 'Know how to solve every problem that has been solved.'[1]

Ramin Hasani, a postdoc researcher at MIT, quoted Feynman's first saying to me when describing his pioneering work in the field of artificial intelligence (AI). His research is a wonderful example of the connection between observation, insight and discovery. It demonstrates how scientists can shed light on complex subjects by breaking them down into something that is both easier to study and observe, and more adaptable is the face of new observations. My personal holy grail.

Hasani's subject is neural networks: computer programs comprising layers of nodes that replicate the function of neurons (nerve cells) in the brain, firing messages between each other to interpret and sort data sets they have been trained to scrutinize. They are often used to tell the difference between things, like whether body tissue is healthy or cancerous, or whether a message has been written by a human or a computer. The concept is actually quite an old one, first proposed theoretically in 1944, and used all around us today: underpinning the spam filter on your inbox and the instant checking of whether you qualify for a loan or overdraft. It's why, even on a particularly bad-hair day, the facial recognition system on your iPhone will still love you.

As these examples suggest, neural networks are pretty good at telling the difference between things, identifying patterns and sorting them into groups. It gets more complicated when we ask AI to try and do things that the human brain takes for granted, and which we loosely define with words like intelligence, intuition and consciousness. This was the problem that Hasani, like so many AI researchers, was starting to explore back in 2014. He was 'trying to figure out what is the role of randomness in communication between islands of neurons' – in other words to map how a brain actually works, trying to understand what he wanted to (algorithmically) create.[2]

But unsurprisingly (after all, the human brain has 86 *billion* neurons working in it), this quickly left him with more questions than answers. '[The] more I got to know, [the more I] realized we do not have a lot of data about these things, and it's really complex to even start looking into the question of what is intelligence.' So he changed tack, looking for a different lens through which to observe – one that would allow him to see more clearly. If a complex system like the human brain was too hard to understand in full detail, perhaps a less sophisticated one would do. Looking for answers about the real-world behaviour of neurons, he decided to seek out the simplest example he could find. 'I said I have to do this principally. Let's find out the smallest nervous system that is available in the animal domain . . . this was the brain of the nematode, the worm. This worm has 302 neurons, [a] super-small nervous system, but it's still a sea of information. It's crazy how much we still don't know about that small brain.'[3]

The simplicity of the worm's nervous system allowed it to be studied and understood – to be observed – in a way that would not have been possible with a more sophisticated animal. Hasani recalls in our interview: 'I had to go from the nervous system of the worm, to go deeper and deeper, to understand how do neurocircuits, fundamental circuits from the nervous system that perform a function, how do they work? And then in order to understand that, you have to understand the basic mechanisms of how neurons and synapses [pathways through which neurons send messages] interact with each other.' The outcome of this work was a discovery that both delighted and surprised him. 'I observed that a network of 19 neurons with around 200 synapses can drive a car. That was shocking. That was the part where we got extremely excited . . . because we saw with super small capacity, we could create amazing dynamics.'

Hasani's observation laid the foundation for the development at MIT of what are known as 'liquid' neural networks. In contrast to their traditional cousins that are essentially preprogrammed, trained to classify information based on what they have 'learned' from a vast database of previous examples, the liquid networks are designed to be adaptable. In effect they learn on the job, changing their behaviour based on new inputs rather than seeking to fit them into a pre-set framework. As Hasani explains, this is a crucial step towards AI that can replicate some of the more sophisticated functions of animal thought processes. 'The real world is all about sequences. Even our perception – you're not perceiving images, you're perceiving sequences of images.'[4] The greater sophistication of liquid networks could significantly improve the application of AI in real-world settings such as self-driving cars and automated devices in factories or hospitals, where the ability to react to changing or unforeseen circumstances, including human failure, is fundamental to risk management. Their adaptive properties represent another step towards the white whale of AI: replication of the full flexibility, creativity and intuition of the human brain – artificial general intelligence.

The liquid networks are not just a significant development in the field of AI, but a counter-intuitive one. Any researcher would assume that a better algorithm, capable of greater agility and more advanced interactions with data, must necessarily be a more complex and powerful one. To replicate the human brain, it would seem obvious that you need an AI with equivalent bandwidth.

Yet the MIT team showed that a step towards this kind of

advanced functionality was possible based on a neural network that was almost absurdly small, with a fraction of the equivalent capacity that a worm uses to wriggle around. By observing the simplest and least sophisticated nervous system they could find, and understanding it completely, they had isolated features that could take AI another step closer to replicating the most sophisticated. They achieved complexity by starting with simplicity: narrowing their area of study down to a canvas on which they could isolate key behaviours and characteristics. And in the process, they had harnessed one of the eternal truths of science, that discoveries generally begin through observation – detecting trends, noticing anomalies and identifying what may be transferable from one context to another. The most advanced science in the world often starts with the simplest and most fundamental of human skills: to open our eyes, observe what is happening around us, and question what we see.

The idea that science begins with observation is straightforward enough. Before there can be a thought, forming a hypothesis that shapes an experiment that may prompt a discovery, there must be an observation – the recognition that something is either suitably interesting, different, wrong, unusual or unexpected to merit further investigation.

Observation also presents a paradox: incredibly important but, at least in conceptual terms, incredibly simplistic – so straightforward that a child can do it, as of course children constantly do. Many of science's most important discoveries may have started with chance observations, but scientists are not necessarily ready to credit observation with the same significance as theorizing, experimentation or peer review. As the historian of science Professor Lorraine Daston has written, observation is 'ubiquitous as an essential scientific practice in all the empirical sciences' but also 'invisible because it is generally conceived to be so basic as to merit no particular historical or philosophical attention'.[5]

By failing to take observation seriously, we do both it and ourselves a disservice. Observation is not just essential to any scientific process. It is also one of the things that humans are uniquely well equipped to do with a high level of creativity and competence. While it is true that machine intelligence plays a growing role in scientific discovery, doing the heavy lifting of carving out patterns and anomalies from vast swathes of information, it is still very limited in function. So next time your phone decides something for you, be a rebel and don't hesitate to question it.

Consider a piece of advice I like from the statistician David Spiegelhalter: 'Stick to what you know about & shut up about everything else.'[6] This is helpful wisdom for people about to publish their next Twitter (now rebranded as 'X') thread, but a sentiment that algorithms take too far. They mind their own business so much, a logic encoded by design, that they are incapable of seeing beyond the parameters that have been programmed. Even with advances like the liquid algorithm, AI cannot yet match the intuitive beauty of the human brain, with its ability to manage contextual fluidity, subconsciously clock information and tuck it away for later use, or see something that a model would deem normal but which instinct suggests may not be. Against this, AI still faces what has been described as the 'binding problem': it can process and store vast quantities of data and detect patterns within it, but also struggles to knit together the information it encounters intuitively, matching the human ability to 'understand novel contexts in terms of known concepts, and therefore leverage . . . existing knowledge in near-infinite ways'.[7] Whereas human thought, strong on cognitive flexibility but weak on raw processing power, relies on the ability to conceptualize as a route to understanding, oppositely equipped AIs 'mostly learn about surface

17

statistics in place of the underlying concepts, which prevents them from generalizing systematically'.[8]

These are some of the key skills of observation in the scientific context: the ability to see things we can't yet quantify, notice things that will be useful once we understand them better, and detect the shadow of a problem or inconsistency that is begging to be probed further. Observation means peering into the mist to estimate the gap between what we already know and what we would like to know. It sets the frame for what comes next, directing but not restricting the subsequent stages, a little like the colours on a painter's palette.

Observation isn't just intrinsic to all scientific work. Often, it's the *only* tool available to scientists who are working at the cutting edge of a new field or trying to grapple with an emerging problem. As the whole world discovered during the Covid-19 pandemic, that is especially true of epidemiology – the study of diseases and how they emerge and spread in particular populations. Many of the basic tools that scientists used to track the spread of Covid were in principle little different from the techniques pioneered by Dr John Snow, who worked principally as an anaesthetist, over 160 years earlier. Studying a cholera outbreak in London in 1854, his mapping by hand of cases and deaths identified the source as a single contaminated water pump on a Soho street. By observing that some populations in the affected local area had remained free of disease, including workers in a nearby brewery who had access to their own water supply, he was able to establish that the critical factor was not common air but shared water, contradicting the prevailing scientific belief of the time that cholera was an airborne disease. By observing what he could actually see, rather than what established theory suggested he should, Snow was able to question the received wisdom and ultimately make a critically important breakthrough. During the Covid pandemic we all became amateur epidemiologists, tracking data dashboards and learning what the R (replication) number meant for the spread of the virus. It showed that, while tools and techniques may have moved on since Snow's day, the basic principle of observing what is actually happening, rather than making assumptions based on what you expect, holds true.

The power of observation in science becomes clearer when we start to consider that it is not actually a straightforward process at all, even though the principle is far from complicated. One challenge is to come to terms with the limitations of what we are observing, which is usually only a small fragment or subset of an entire system, indicating something that may be a trend or could be an anomaly. Another is that we are incapable of observing anything in complete isolation, without a whole host of preconceptions or expectations. The way we see ourselves in the mirror is likely to depend not just on the image that reflects back at us, but how we are feeling that day, who we have spoken to that week, whether we've been for a run recently, and if we're wearing clothes in our favourite colour.

In the scientific arena, researchers are invariably not just looking *at* something, but looking *for* something. They start with a question, and sometimes a hypothesis (or even just an instinct), in mind. John Snow was seeking the source of an outbreak of deadly disease, the latest in a recurring series of cholera epidemics. Ramin Hasani was looking for answers about how exactly the nerve centres and pathways in an animal's brain work, and whether they conform to patterns that can be replicated. Not every act of observation is so directive, but it is selective in any case – either through the ideas or assumptions you start with, or the possibilities you discount from the very beginning. This process of selection, the intuitive ranking of the evidence in front of us, is an important part of any scientific undertaking, even as it undermines the notion that there is such a thing as objective truth or unbiased research. 'The truth is that this is a world in which there is an infinite number of facts available, and we have to begin with some feeling for which are important and which are not,' the academics Inge and Martin Goldstein suggested, channelling the dictum of Charles Darwin: 'How odd it is that anyone should not see that all observation must be for or against some view if it is to be of any service!'[9]

The preconceptions that scientists bring to the process of observation can, of course, frustrate as well as aid effective research (and the same is true of algorithms that, far from being neutral robots, carry the biases of the humans who developed them). What might be a shortcut to a meaningful hypothesis or discovery can also be a fast-track to missing the point altogether – trying to hammer observed evidence into the form of a theory it does not support.

At the same time as Dr Snow was ending the 1854 cholera outbreak by having the contaminated water pump shut down, the government's Board of Health was beginning an inquiry that would eventually dispute his findings – not because it had superior evidence, but simply because it had failed to properly investigate the issue of waterborne transmission in the first place. 'They did not ask which pump the victims had used, nor did they investigate whether the pumps could have been contaminated from elsewhere.'[10]

As well as demonstrating how entrenched an established consensus can be in science – the view that cholera was transmitted through air rather than water continued to dominate until another major outbreak twelve years later – this episode illustrates both how scientific observation can go right and where it goes wrong. Scientists need to have hypotheses and preconceptions to guide them towards observing what is relevant, otherwise they simply become lost in the mass of potential data, unsortable even with the benefit of modern computing power. They must use their training and experience to give shape and meaning to an observation. Yet if they are too directive and cling to preconceived ideas too tightly, they risk discounting alternative explanations or possibilities before they have been properly considered. This balancing act, being neither so open-minded as to be directionless, nor so close-minded as to prejudice the inquiry, is a constant challenge of any scientific undertaking. It's the sort of dilemma I find myself in every day trying to navigate my ASD brain through London, where a million sensory stimuli lurk around every corner and small decisions like how to cross the road or where to get my morning coffee can become agonizing. I should know from past experience what to do, but sometimes still get lost in the constant distractions and endless options, leaving me standing on the kerb for minutes at a time, sadly without caffeine. Try justifying being late for the lab to the vice president of UCL purely because you were too busy contemplating your senses on your morning commute. Very bougie.

The question of deciding what to observe is further complicated by whether scientists can always trust in what they think they are seeing. In 2011, researchers in Italy reported an astonishing discovery. They had recorded that neutrinos (subatomic particles) beamed from a particle accelerator at CERN (of Large Hadron Collider fame) arrived at their location, around 730 kilometres away, faster than the speed of light – a race they won by a whopping 60 nanoseconds. You read that right: they'd broken the speed of light! This was a remarkable finding, one that promised to change the face of modern physics, contradicting as it did Einstein's theory of special relativity, which established the speed of light as a constant for any object in the universe, something that cannot be exceeded. And it was not some crazy one-off: the experiment was run around 15,000 times across two separate, published studies. While some observers immediately discounted the findings, others suggested that they might indeed be possible. If these particles were somehow travelling through dimensions outside the four (three of space and one of time) that are part of the established physical universe, it was feasible that they had indeed arrived faster without actually breaching the speed of light as Einstein had defined it.[11] However, further studies revealed a more prosaic reality, with a fourfold experiment showing that neutrinos were travelling within the speed of light in each case. Rather than any miracle or new mystery of physics, it transpired that the initial findings had been impaired by two simple equipment issues: a faulty cable and a measurement clock that was running too fast. What had initially threatened to overturn more than a century of theory ultimately reinforced something that every scientist knows, from the chemistry lab at school to the most advanced facilities in the world: how you measure something is as important as what you discover. A scientific discovery is only as valuable as the method that underpins it, and the metrics used to define it.

Observation may be one of the oldest, most fundamental parts of scientific study, but that does not mean it stands still. The importance of observation is perennial, but the tools at our disposal to undertake it are constantly changing and improving. Sequencing the human genome, a process that cost a neat $300 million when it was first done in 2000, can be done by almost any lab today for just $1,000.[12] It's a similar story right across the spectrum of scientific research: tools and techniques that would once have been unthinkable or prohibitively expensive are increasingly available and widely used. As important a benefit as this can be, it also creates its own problems and poses new questions. Just because we can crunch more data or call on more experimentation, does that mean we should? Is more data always better, or is there a point at which it confuses more than it clarifies? And what role remains for human intuition in detecting patterns, distinguishing signal from noise, and deciding how to shape the reams of information that can now be accessed at the click of a button? Is there more than one way to 'bind'?

Thanks to our increasingly technological lives, we face a similar oversupply of data outside the laboratory. It's all too easy to let your life become defined by the number of steps you take in a day, how many hours' sleep you got the night before, how many tries you took on Wordle last week, and how much extra time you spent working from home. All carefully catalogued by the smartphone that will then scold you for your excessive screen time. Measuring our lives in this way is a double-edged sword. We can become so obsessed with plotting the coordinates of our daily lives that we start to forget what really matters. This is what I think of as the Fitbit effect, where quantification offers the false promise of validation. Did we really cover all those kilometres if we didn't measure them? It's not hard to see the problems this can lead to, if someone becomes so focused on tracking their life in data that they start to forget about the actual living part and see what is in front of them – the opposite of what being fluid, like the algorithm, is about.

There are clear limitations to what observing and gathering data about our lives can achieve: you're probably not going to be happier if you start counting up how many times you laughed or smiled in a day, for example. We are encouraged more and more to track our lives like science projects, but – especially for minds that tend towards the obsessive – this risks doing more harm than good. It's the mindset of the person who is so busy taking snaps of a social gathering or a brilliant meal that they forget simply to enjoy the moment. And even if all this memory clutching and data gathering doesn't stop you from having fun, it may not actually reveal that much. As Albert Einstein said, in one of his lectures on geometry (and one of my favourite scientific quotes): 'As far as the laws of mathematics refer to reality, they are not certain; and as far as they are certain, they do not refer to reality.' It's a good reminder that there are both limits and costs to quantification, and as much as I live for the feeling of closure and clarity, I have to remind myself that exactness doesn't completely map onto trueness. Always ask yourself why you are observing or measuring some part of your life: is it helping you towards a concrete goal, like building up your distances towards running a marathon, or is it just a kind of crutch or habit that you could actually do without?

It's the same in science: in any research project, there comes a point when you need to move from looking to doing, and to use everything you have learned to build and design the solution (or part of one) for the problem you initially set out to solve. When Ramin Hasani was studying the nervous system of the nematode, he drew a line past which further observation would not have served the best interests of his project. 'We got into the neural circuit level, and we went into the neuron and synapse level to really fundamentally figure out what the building blocks are. And you . . . can even go lower than that and computationally model it down to atoms.' But there is actually a level that you have to satisfy yourself that you don't want to go below, in order to . . . take this model and see what capabilities you have.'[13]

As he suggests, sometimes you have to stop observing to allow you to make progress with an inquiry. But in other contexts, patience is required. Wait long enough and evolutions in scientific capability may transform your ability to observe a problem. To take the example of cancer research, a field in which I have worked as a bioinformatician, improved technologies in areas from genomic sequencing to modelling the 3D structures of proteins are revolutionizing the scope of research. These don't just produce more data, but layer it together in new ways to offer fresh context and perspective.

Where researchers once relied exclusively on animal models, approximations and potentially unrepresentative tumour samples from biopsies, they are now able to reconstruct and study different cancers in both a more realistic and a more dynamic way: turning what were effectively grainy snapshots into a much more complete and lifelike picture that can be studied from all angles. 'There's been a lot of time in cancer research where people have worked on mouse models,' Professor Frances Balkwill, lead researcher on the CanBuild project into tumour microenvironments (the complex of cells and structures in the body that surrounds a tumour), has suggested. 'One of the mottos of my lab is that if you're a human being with a mouse cancer, we can help you.'[14]

Her research programme originated, like so many, with the observation of a clinical problem that was equal parts perplexing and intriguing. As she told me: 'The aim of the grant is to increase survival of women with high-grade serious ovarian cancer. That is an interesting scientific problem, because a majority of these women respond to chemotherapy very well and seem to be cured, and then they relapse. What we and others know is that the chemotherapy stimulates the tumour microenvironment, stimulates some sort of immune reaction, but it's never good enough to last.'[15] This clinical observation helped to reinforce another – that the prevailing methods of studying such tumours were insufficient for the level of understanding needed. Tumours were so frequently relapsing or resisting treatment over the long term that the scientific understanding of them must be missing something. Moreover, advances in tissue engineering meant that it was becoming possible to move beyond a status quo in which cancer cells were studied as static entities, in isolation and out of context. 'What we routinely have done for many, many years, and deep inside me I have always felt was wrong, [is] we've grown our cancer cells on basically a special type of plastic. But the body is not plastic,' Professor Balkwill said at the outset of the project in 2013. The goal of CanBuild, as she defined it, was much more ambitious: 'To reconstruct a human tumour that would grow and evolve as a human tumour might in the body.'[16] In other words, the project would equip cancer researchers with a new way to observe and study the problem.

The research she has led at Barts Cancer Institute has done much to overcome old limitations. CanBuild has focused on studying not just ovarian cancer cells in isolation, but what is known as the tumour microenvironment (TME) – everything that surrounds the tumour and which it can over the course of its growth corrupt, including the cells that comprise the body's vascular, nervous and immune systems – essentially the context in which the tumour lives and grows, and off which it feeds. These constituents of a healthily functioning human body have tended not to be the focus of cancer research, even though they can make up a significant part of a developed tumour: from half its mass to up to 80 per cent in the case of pancreatic cancer.[17] It is the body's infrastructure that cancer cells hijack for the purpose of their growth; a process that is as important to understand as the nature of the malignant cells themselves. While much of our propensity to particular types of cancer is genetic, the body itself also plays a significant role in determining the level of risk. As Professor Balkwill has suggested: 'The genetic damage is the match that lights the fire, but the tumour microenvironment is the fuel that fans the flames.'[18]

Over the last decade, her project has both achieved its proximate goal of re-creating the tumour microenvironment, and gone a long way towards identifying patterns of interactions between cancer cells and the TME which could be the basis for future treatments. Improving technology has enabled CanBuild to deconstruct tumour samples taken from patient biopsies and reconstruct them, piece by piece. It first grew artificial cells that replicated the omentum (fatty cells of the ovary) in which these tumours typically develop, before adding the cancer cells themselves (harvested from real tumours) to observe how they grew and interacted with this simulation of their microenvironment. Think of it like SimCity for cancer research. From that painstaking, ground-up process of deconstruction and reconstruction, creating mini-tumours in petri dishes, a number of significant discoveries have followed. One was that platelets (the blood cells that perform a clotting function to prevent excess bleeding) play an important role in helping cancer cells to spread, by breaking down the protective layer of mesothelial cells around the surrounding body tissue. Another was to isolate a part of the body's immune system, a cytokine (protein produced by immune cells) known as TGFβ, as one of the key enablers for cancer cells.[19]

These might sound like relatively small details, but they are critical insight for researchers who are looking for new ways to target and treat cancers that have proven stubbornly resilient in the face of current medicine. Any tumour is the product of a multiplicity of cell structures, molecules and chemical messengers, enabling and facilitating the growth of cancer cells through previously healthy parts of the body's main organs. Like a game of Jenga, one goal of cancer research is to establish which of these bricks, if removed, could bring the entire edifice crumbling down. Or, in other words, where treatments could most effectively be targeted to prevent such tumours from taking hold and metastasizing in the first place. The better we understand the biological behaviour that underpins different types of cancer evolution, the more effectively we can ultimately develop countermeasures that are individually tailored to their growth patterns and most dangerous bodily enablers.

Which brings us back to the importance of observation – not just what is observed but how. The real terror of cancer is how dynamic an entity it is, one that can grow through our bodies with devastating speed as it co-opts and corrupts the very systems that work to keep us alive in the first place. The beauty of a project such as CanBuild is that it takes an equally dynamic approach to observing the behaviour of tumours. By re-creating the best lab-grown proxy of an actual tumour that has yet been achieved, it allows researchers to throw all the data-science capabilities of modern computing not at a still image but at a moving picture – one with texture and dimensionality, and from which some of the more subtle behaviours and characteristics of the tumour can be deduced. A dynamic approach to observing dynamic systems is already taking us closer to answering the perennial question of how the body can be set up to fight one of its most deadly assailants.

In parallel, such advances are also a sobering reminder of how much we do not know, even about a subject that has been as extensively researched as cancer. The better our tools become, the more clearly we can see what we have been trying to observe, and the more new avenues of research and debate open up in front of us. As one scientist I interviewed for this book put it, the research process can be likened to stumbling about in an unlit room looking for the door, which you finally locate only to find yourself in another, bigger, even darker room. Both the frustration and the fascination of the job for any scientist is that your ability to provide one answer invariably creates two or three new questions in turn. The better and more accurately we are able to observe something, the more it demands to be studied in new and different lights.

Whether we realize it or not, we are spending our whole lives observing and taking in the world around us, be it a smoky city or a sleepy village. Like the neural network that has been trained with vast databases of information, our brains use our aggregated experiences to process everything we see, hear, smell, touch and taste – flashing instant messages to the toe that needs to be removed from too-hot bathwater, or the frontal lobe where decisions are made about when to investigate an unusual smell or move away from a potentially dangerous situation. These subconscious, effectively automatic observations and decisions are happening in parallel with those we notice ourselves seeing, like whether a colleague has a new haircut or the orchid in your kitchen has started to flower (or in my case to wilt from over-watering). Sometimes, the decisions we think we are making consciously have already happened in the subconscious realm by the time we catch up with them. Researchers at Princeton in 2006 found that subjects who were given only a tenth of a second to judge someone from their facial expression reached similar conclusions as those who had unlimited time – especially when it came to deciding about their trustworthiness.[20] I often use this flashcard method when I want to tune in to my gut feeling at a given moment, where I ask myself: 'What is my environment telling me and how do I feel about it?' in no more than ten seconds.

We might not all be scientists, but we can easily be more scientific in how we observe and navigate the world around us. That starts with being a critical observer of yourself, recognizing what you tend to pay attention to and why. Like a scientific experiment that is based on bad data or skewed parameters, we are going to get a limited perspective if we only trust news articles from sources to which we are politically aligned, or take feedback at work seriously only from people we like. We need to recognize that we are never observing anything with truly fresh eyes or free from a boatload of preconceptions, expectations, hopes and frustrations. The space between observation and decision making is often so small as to hardly exist at all. We have, in effect, decided before we've really had the chance to observe and think.

The lesson from science is not to try and rid ourselves of these preconceptions, but to adjust for them. We are immutably the people and personalities that our unique experience of life has formed. But that does not mean we have to be close-minded and unwilling to consider alternative explanations to the one we already had in mind. Once in a while, the outcome you had discounted as an impossibility is actually going to happen, and the dead cert won't come to pass. Occasionally, whether you like it or not, the worst person you know really is going to make a good point.

Good researchers are prepared to walk this line between fixity and flexibility. They neither start with an empty mind, willing to be taken wherever the observation directs them, nor do they stop allowing that they may yet be proven either wrong or misguided. As we will explore in the next chapter, sometimes the most worthwhile experiments emerge from studies that began with a different objective or were focused on another side of the same problem. The thing you initially thought was interesting might prove to be a dead end (like a crush that quickly fizzles out), but one of the by-products of that research could be a goldmine of further work and future discovery: when we learn about the people who aren't good for us, it helps with finding those who really are. It is a confident open-mindedness, steering in a certain direction while being willing to change course when needed, that stands the best chance of taking you somewhere interesting. We need, like the liquid algorithm, to be fluid. This means you benefit from the power of observation without being overwhelmed by the breadth of possibilities that an entirely open mind would entail, yet being mindful of how much attention we are giving to a certain decision.

Scientists show us the power of observation that is directive without being prescriptive. They also remind us that we need to be honest with ourselves about what we are observing, and wherever possible to see things in their proper context. As the CanBuild programme has demonstrated, there is a world of difference between looking at a cancer cell in isolation and studying the entire tumour microenvironment – observing not just the relevant cell structures as static entities in isolation, but how they interact and behave in the dynamic environment that is their reality. If we are being honest with ourselves, most of us probably don't apply this principle of context enough in our lives. Do we really stop to think why a friend or colleague who has been unusually abrupt with us reacted that way? (I must admit, this is where my imposter syndrome comes in handy as I am always thinking about context and where I might go wrong.) Do we get offended and annoyed, and decide to send the passive-aggressive email we have typed out, or do we stop to think or ask what particular pressures they may be facing in their micro-environment? And with ourselves, when we are having a bad day or struggling to get something done, are we more likely to bemoan our lack of productivity or be honest about the reason things aren't quite right? Science reminds us that looking directly at something isn't always the best way to observe, especially if this limits our ability to see what is going on around it. Sometimes we shouldn't take a friend's reassurance that they are 'fine, really' for granted, and should prod them to tell us how they actually feel. With the scientific mindset, you don't just accept the first thing you see or hear: you question whether it can or should be trusted, and what it is really telling you. You consider your own point of view, that of the person you are talking to, and how both might impact what is being said and understood. When it comes to people talking about their mood, the context (body language) often says much more than the words coming out of someone's mouth.

This is a reminder that observation is not the same as looking. Fascinating and stimulating as it can be, we are doing it to learn something and not simply to admire the view. To serve a purpose it must have a direction and an end point – by which we have learned enough and picked up enough threads to start stitching a hypothesis, devising an idea or coding an algorithm. Much as Darwin argued that observation 'must be for or against some view', it also needs to have an end in mind to be useful. Of course a scientist never actually stops observing, learning or questioning what they think they have discovered. But nor do they delay the point at which the seeds of initial observation must be harvested to grow into something much greater, continuing the process by which a notable observation can ultimately become a meaningful discovery.

2. Hypothesis

How to come up with ideas

A lot of the questions I ask myself begin with 'What if?' Some are in my control (What if I took a different route to work? What if it was a sandwich day and not a salad day?) and others are not (What if none of the next six cars to pass the crossing are blue?) And what if all of these things happened or none of them did – would it matter or make a difference? Some days I make a sport of contradicting my normal routines just to see what is and isn't important – to take myself to the point of maximum contrast between perception and reality, where the soft white of a fried egg meets the crispy edges.

I play games like this in my head, and have done for as long as I can remember. I do it for fun, out of habit, and because a part of me still hopes that I can sniff out some kind of pattern or meaning in even the most mundane parts of my life. And I keep on doing this because it is, at the most basic level, scientific: the desire to connect things that hardly look like dots, to build bridges that might fall down, and just to try out ideas that sound a little wacky. This is the next stage in any scientific inquiry, taking what you have observed and trying to make sense of it, asking if maybe this is how it works, and what if this was the explanation for that. As scientists we have an impressive-sounding word for all this wondering and what-iffing: the hypothesis.

When I spoke to scientists about the role of the hypothesis in their work, some of the responses surprised me. Two researchers, from fields as varied as psychology and theoretical physics, used the same metaphor. They talked about something seemingly unscientific – about the power of stories. 'We make observations, we see associations, but we need to put them into a mechanistic framework, into a story basically,' said the neuroscience researcher Dr Katharina Schmack, an expert in the study of psychosis. 'I call it a story because it's a narrative in the end. And this leads us into a testable hypothesis.'[1] That thought was echoed by the physicist Dr Chiara Marletto. 'As a child I liked crime fiction stories,' she told me. 'It really feels a bit like that, because you've got these clues you need to put together but you don't see the full solution most times, and when you have a setback, it means you haven't put them together correctly and you have to put them back on the desk and think about them again, in perhaps a more creative way.'[2]

I love this idea of a hypothesis being like the premise for an unputdownable thriller. It sounds like the opposite of real science: emotion and subjectivity in the place of logic and reason. Except that isn't really science at all, or at least not the whole of it. Despite its reputation for precision, order and certainty, scientific research is often the very opposite of those things. Far from being a perfectly ordered world of infallible theory and error-free execution, a lab is actually a place where chaos and confusion often reign. Much of the time, it is a maelstrom of coffee-stained scraps of paper, out-of-control whiteboard scribblings and data points that appear to make no sense at all. It's only after a huge amount of head-scratching, trial and error, and squinting from different angles that answers start to emerge from the morass. And, of course, these answers immediately pose a dozen new questions, sending you back to that very busy drawing board. You must constantly play devil's advocate to the same work you are trying to advance and verify.

As Dr Schmack put it to me, the way scientific research is

presented and communicated can reinforce simplistic perceptions about how its findings were achieved. 'We read [scientific] papers and the story is always quite linear: we had a hypothesis, we tested the hypothesis, we found this, we made that conclusion. But in reality science is very often not like this. We make one observation then come up with a hypothesis, then we go back and we modify. It's much more chaotic than the linear story that gets told at the end.'

A recognition of this chaos is essential to understanding the hypothesis, a theory or explanation that is yet to be proven and one of the most important and curious beasts in science. This is what every research effort rests upon. In many ways it is the foundation stone of further investigation – something to build upon and sharpen ideas against. But the hypothesis is also a fluid, uncertain thing, liable to change and subject to being revisited when new evidence emerges. It is a conjecture, more a weather vane pointing you in different directions at different times than a reliable true north. The hypothesis is a beautiful reflection of the messy, infuriating and occasionally spellbinding nature of the scientific process. It highlights the circularity of an endeavour that so often takes you back to the point where you started to check assumptions and consider new approaches. And it reinforces the precarity of any research process – which may fall apart and need to be rebuilt for so many different reasons: perhaps you were looking for the wrong thing, in the wrong place, or using the wrong method.

For any scientist, the hypothesis is as humbling as it is necessary. It symbolizes the uncertainty you are dancing with, a reminder that this is a world composed as much of ifs and maybes as it is of proofs and theories. The hypothesis is a bridge you build, in full knowledge that it might collapse as you try to walk your ideas across it. The possibility of such failure is intrinsic to the process, and of course becomes more likely the more you experiment and put yourself out there. In this way a hypothesis is a bit like a job application by another name. It's an idea you are trying to make real, a ledge you can see but don't know if you will be able to reach.

Just like with the application, you have to risk failure to get anywhere. As the philosopher Karl Popper argued in the 1960s, a hypothesis is not scientific at all if it is not capable of being tested and proven false. Every scientist who pens one, therefore, has to consider the potential death of their idea at the same time as they are trying to give life to it. This act of reconfiguring can often feel demotivating, unfashionable, and hard to engage with, especially if it brings us 'off track'. But part of building a theory is reshaping it, and not just once, but *all* of the time. To the point where my PhD supervisor said in our first meeting that two of the most important parts of research are i) having a plan and ii) not being afraid to throw it in the bin the next day. Hearing this was something that I definitely needed (along with a thick block of Post-it notes). Especially as a bright-eyed, perfectionist graduate student with an enervating need for routine.

The intrinsic mortality of a hypothesis might sound pretty wobbly and scary (or as a scientist would say, great fun), but that doesn't mean hypotheses are something we can't rely on or draw firm lessons from. In fact, the hypothesis is one of the best ways to understand the scientific process and its virtues. It helps to unlock the puzzle of how scientists create process from chaos and ultimately succeed (in small ways, on *very* good days) in turning questions into answers. It reveals how scientists get their ideas, how they alter them over time, and the contrasting roles of carefully honed theories and chance discoveries. It shows how these both make significant leaps of logic, and reserve the right to change course. Ultimately, the hypothesis gets to the core of the scientific apple: findings that may prove your theory wrong are as important as revealing those which may confirm it. You have to make space for new ways of thinking about or going about a task, even when those contradict something you thought you had already established. Part of you is willing to be proved right, but another, more masochistic side is almost wishing that you will not be (the inner contrarian that is part of every scientist) – because the consequences of a good hypothesis turning out to be wrong could be even bigger and more interesting than the idea you started with. Just as one failed job application (or grant application if that is your gravy) might lead you to be more focused and better prepared for the next one, having honed your ideas and freshened up your skills. It's why I used to have Post-it notes on my wardrobe mirror onto which I had scribbled: 'You are allowed to change your mind, scientists do it every day' and: 'It doesn't make you scatty, unreasonable or unscientific, just flexible and in tune with what is needed.' (FYI this did not involve me talking to myself in the mirror at any point.)

In this, the hypothesis reflects one of the great and most transferable lessons of science – that your requirement to think confidently must be matched by a willingness to discard predictions when the evidence points in a different direction. You can date a good idea and a well-founded hunch, but you mustn't get married to them. And when it comes to forming opinions about pretty much anything, that's a good lesson to keep in mind. Having the humility to walk away from something that has been proven wrong is as important as the ability to develop interesting ideas in the first place. We should be open to giving up strong opinions when the evidence changes, and to letting people change our mind if they have a better perspective or argument than we do. There is nothing to gain from going about your life with fixed views that will never change (apart from a quick ego boost), and everything to admire about people who consider their opinions and outlooks to be hypotheses – ones that need to be tested and validated through evidence, and which may ultimately be proven unsustainable. There is a reason why irony and dark humour are the staple of any science lab: research is an ego-battering endeavour, one that is constantly taunting you with reminders of past mistakes, proof of wasted time, and indications of what you don't yet understand. The hypothesis – a hunch that you both stake everything on and must be ready to abandon at any time – sums up the world of glorious uncertainty that is the scientific process.

One of the most unfortunate characteristics of cancer is its tendency to recur, and to become less clinically treatable when it does so. Many people will be sadly familiar with the reality that, when a cancer comes back, it is usually more dangerous than when it first arose. The evolutionary agility of cancer cells means that they are capable of developing resistance to the drugs that were used against them, neutralizing previously effective treatments. Cancer is not a dumb adversary but a highly adaptive quantity, whose innate ability to evolve in the face of threats makes it so challenging to combat.

Dr Charles Swanton of the Francis Crick Institute has been leading research teams to explore the evolutionary properties and drug resistance of lung cancer for the last fifteen years. His work is changing how we understand the behaviour of cancer cells in the body and, ultimately, can think about devising treatments to match the adaptive intelligence of the tumours themselves. With evolution as its subject, his work has also served to underline the evolutionary nature of scientific research in its own right – how it can emerge out of the fault lines in an existing consensus, grow under the wing of a well-founded hypothesis, and draw strength from chance discoveries and unplanned diversions.

As he told me, the study had its origin in what seemed an obvious discord between the consensus medical science around cancer development and what clinicians were themselves observing. 'When we were at medical school, we were taught that cancers evolve in a relatively linear fashion, and that all cancer cells in a tumour should be approximately the same. Now if that were the case, then one would imagine that resistance would be relatively hard to acquire, when in actual fact it's almost a certainty.'

Based on the consistent clinical evidence of tumours developing resistance to drug treatments they had not initially demonstrated, he and others began to develop an alternative hypothesis: the growth of a typical tumour was following not a linear pattern in which new cells match the old, but a branched one, through which the same ancestor cell produces a variety of offshoots which follow their own different paths and develop their own characteristics. Branched evolution would explain how the tumour that is initially vulnerable to being targeted in a certain way can later develop a form of immunity to the same treatment through the trick of diversity, the official term being 'Intratumour Heterogeneity'. Like all good scientific hunches, this started as just a convincing theory – a potential dotted line between two fixed points, but one that would need to be tested again and again before it could be inked in. 'The way I explained drug resistance to patients in the early 2000s was that their tumours were evolving in a Darwinian manner, but we didn't have much evidence for it,' Dr Swanton recalls.[3]

That insight is typical of how scientists across many different fields begin to form ideas and develop hypotheses. They find fault and take issue with a piece of consensus knowledge, because it somehow conflicts with or fails to explain what they are observing. These fault lines might be small or large, but they create an itch that demands to be scratched through further investigation and inquiry. Like an engine that is making strange knocking sounds, a theory that doesn't fully satisfy the evidence demands to be stripped down and have its parts inspected, to establish whether just a few things need to be replaced, or the entire unit has to be rebuilt.

Scientists are by their nature deeply curious people, who love to question, pick holes in an argument and hold an idea up to the light to see what it really looks like from all sides. (We're also extremely good fun at parties; being used to funding constraints means you can do a lot with a bag of Doritos.) It's this innate restlessness that helps to spur the beginnings of a new idea: a tendency to challenge what you have always been told and to identify even small fault lines in the logic of an established consensus. This is a compulsion which can become self-fulfilling, almost addictive, because, just as the imperfections of any object or being become clearer the more closely you look at them, so the further you examine a textbook theory or principle, the more potential issues you are likely to perceive.

Many of the scientists I spoke to summed up their approach in similar terms. They begin with a problem, something they either can't explain or really want to explain, a potential contradiction to the consensus or a missing piece in a chain of evidence. The question that has been bothering them must then be formed into a hypothesis that can be tested until it is proven or disproven. That almost starts to sound neat and tidy until you remember where we started this chapter – with chaos, coffee stains and brain fog. Because, like the evolution of a tumour, the growth of a scientific inquiry is not linear but branched. It does not (unless you're very lucky) follow a predictable sequence of stages that can be taken in calm succession. Much more likely it will zigzag all over the place, squirrelling off in different directions based on what you discover and the sometimes unforeseen implications of those findings. Like in life, where if you are brave enough to deviate from the prescribed path, you earn the space to make your own and colour it how you like. A research project may be something bounded by grants, mandates and red tape (sigh), but it is also an entity that can evolve in unpredictable ways, which is both a glorious and a gloriously frustrating thing.

Charles Swanton's research on cancer evolution was a case in point. His team made a notable breakthrough in 2011 when conducting a detailed study of a kidney tumour, sequencing all of the roughly 22,000 cancer genomes it contained, and finding early indications of the branched-evolution hypothesis: 'There were many more mutations that were different and distinct between regions than . . . were similar.' One such mutation was found in the same region of the tumour targeted by the drug typically used to treat kidney cancer. 'In that one region we could find an activating mutation, but not in other regions of the tumour. So that helped us . . . begin to prove the hypothesis that drug resistance is likely being caused by diversity in the tumour at baseline, but also diversity that evolves over time.'[4]

So far, so logical. Except this important finding arose almost by accident. 'This was actually a control experiment,' he recalls. His grant at the time was concerned with finding what are known as biomarkers (medical indicators) in the tumour, using then cutting-edge genomic sequencing to try and predict how a tumour might respond to popular drug treatments. The experiment described above was seeking to create a baseline for this work – to establish if biopsies could be taken from the tumour that would give a consistent picture of whether biomarkers were present or not. The outcome was a classic case of a door closing and a window opening. 'We had to be sure that wherever we put the biopsy needle [was] going to give us the same result. And, unfortunately, it gave us a very different result depending on where you put the biopsy needle,' he remembers. Yet that failure proved to be a spur towards establishing the branched-evolution hypothesis. The impossibility of getting a consistent sample from different places within the same tumour meant that, by straightforward deduction, this could not be a case of linear evolution. Varying biopsies must mean that the tumour was developing in different ways in different places. The reason for one experiment hitting the buffers provided the first piece of a new and fascinating puzzle: why and how tumours evolve in a heterogeneous way, and the implications of this for treatment. 'We were really unable to find robust biomarkers of response to these drugs, so instead we spent the next ten years studying how or why tumours evolve,' Dr Swanton says.[5]

This underlines the very dynamic way in which a workable hypothesis can emerge – variously from logical leaps made by eyeballing the evidence, dissatisfaction with existing explanations, and conclusions that may fall sideways out of experiments that were initially pointed in a different direction. You can't take it personally when the evidence proves your ideas wrong; in fact, it's not science at all unless you are being guided by what you actually find rather than what you hoped to see. In line with one of my personal PhD sayings: 'If you don't have an identity crisis during your PhD, then you probably aren't doing it right.' In this vein, important scientific discoveries are likely to be some combination of the focused and the accidental, the foreseen and the unforeseen. When you are exploring the unknown, that is simply the nature of the beast. You are working with what you discover along the way as much as the bag you packed at the start. That can be an intimidatingly unstructured way to go, but it is also what makes research so compelling. You never know what you are going to find, but you keep going, because it might be even better than what you had hoped for. The failure of the primary hypothesis does not necessarily bring your work to a shuddering halt. In fact, it may point you in a new and more exciting direction than first envisaged.

Sometimes a hypothesis will be a very specific thing, targeted at a particular problem and with a single test or experiment in mind. As a scientific tool, the hypothesis can also be much more wide-ranging than this, reaching across subjects and disciplines to encompass swathes of knowledge and possibility. This can and does include having a hypothesis for how to make hypotheses (a bit like having a meeting about meetings) and pondering new ways of thinking about how to approach scientific problems from first principles. This is where it *literally* all gets a bit meta: the science of how to do science.

One such hypothesis has been gaining ground in recent years, and it's a big one. So big in fact that it seeks to change the fundamental basis on which physics has worked since the days of Newton and Galileo. Chiara Marletto, who has developed this new theory alongside the physicist and quantum computing expert David Deutsch, explains: 'The deepest question is whether it's possible to find a more fundamental mode of explanation for physical reality than the one we already have. Currently the best you can do if you're a theoretical physicist is to find a dynamical law that can be applied to the whole universe and, given the initial conditions of the universe, it can let you know what the universe will do at some point in the future or in the past.'[6]

In other words, our entire conception of the world is based around what is known as fundamental physics: rules such as Newton's laws of motion and Einstein's theory of relativity, which give us tools to track and predict the movement of objects through space and time. Marletto and Deutsch are not questioning the efficacy of this approach, but they do challenge its ability to provide all the answers about the physical universe that we are seeking. As Marletto has suggested: '[There] are some phenomena in nature that you can't quite capture in terms of trajectories – phenomena like the physics of life or the physics of information.' There is so much about the richness of human and animal life (she gives the example of eyesight, and how the anatomies of the eye 'coordinate exquisitely to permit vision, just like the optical components of a sophisticated camera'[7]) that cannot be explained by laws that show how a ball moves through the air or a planet through its galaxy. These would require an explanation that isn't rooted in the doctrine of how objects conform, but based on the various states they could exhibit; where difference is information that provides a signal in its own right.

To augment fundamental physics, Marletto and Deutsch have proposed what they call Constructor theory, and what she has characterized as 'the science of can and can't'. In place of laws that determine exactly how objects will behave in a given set of circumstances, this seeks to define a broader set of conditions: what is possible for the object to do and what is impossible (a transformation known as a task). In this conception, the 'constructor' is an object or system that enables the task to be performed through the lens of possibility (Can it do something or can't it?) rather than exact position or outcome (What happens here if we do this?).

Like the best physics, this is both simple and profound. There is nothing new or dramatic about exploring the art of the possible or thinking in terms of what the laws of physics deem to be impossible. Yet there *is* something quite dramatic about shifting perspective from tracking what does and does not happen to what could or could not. By moving from the factual to the counterfactual (the expression of which actions are possible or impossible), Constructor theory encourages scientists to unhook themselves from what they can and can't prove to what they can conceive as possible. It overcomes the 'misconception . . . that once you have specified everything that exists in the physical world and what happens to it – all the actual stuff – then you have explained everything that can be explained'.[8] As Marletto posits, we do not need the proof of a shipwreck to explain the purpose and possibility of lifeboats.[9] We must be able to think in terms of possibilities as much as we do tangible, provable realities: the principle of what can happen, and not just the law of what should. Coming from a place of difference, I found this new explanation relieving, and beautifully inclusive, as it provided a place for me to experiment in whatever form I saw fit, and not at the mercy of my peers or what I *should* do – be it how I studied, the way I assembled my dinner from my one-hob kitchen, or how often I took retrospectively cringe selfies in my lab coat purely for giggles.

The authors of Constructor theory believe that their approach could unlock some remarkable new possibilities for the study of physics – unifying physical theories of heat, work, information, and even life itself. The counterfactual approach, Marletto told me, could add dimensions to the science of physics that have so far been considered impossible. 'Traditional physics has adopted a position that [dynamical laws] are the only way to explain scientifically and within physics a given phenomenon. Therefore when something doesn't fit into this mode of explanation, somehow the idea is we should give up on giving an exact physical description of it . . . The typical example is life. We have a good scientific understanding of how living systems work within biology, and physicists would say that an organism is completely compatible with how each of its atoms obey the laws of motion. But we don't hope for laws that actually describe living systems within physics.'[10] Rather than considering some of biology's innate unpredictability as inconvenient or ancillary to the laws of physics, this encourages a more comprehensive approach, encompassing a fuller spectrum of what is possible.

Constructor theory is itself a hypothesis, which hasn't yet produced the new laws of physics that it hopes to give rise to. But even at this early stage in its life, it can teach us something about the power of the hypothetical. The elegance of the theory is in its simplicity: stripping a concept down to its outer limitations under the laws of physics, and trusting in the counterfactual to point towards how a possibility might be achieved and explained. That might sound disconcertingly unmoored from things we can see, touch and prove, but by the same token it may be seen as liberating. Without unhooking ourselves from known quantities and concepts of how things 'should' be, we cannot begin to conjecture about what has not yet been discovered or considered. Approaching new possibilities may sometimes mean departing from existing certainties, even if only to experiment in the theoretical space that has been created. That is good sense for scientific work and not bad advice for life either. Sometimes we have to imagine what we are capable of doing, not just rely on the knowledge of what we have done. We have to see ourselves as constructors with potential to achieve new things, not just repeat the same old ones. Even with a love for routine, it becomes a very timid existence if you rule out the possibility of trying something you have never done before.

All this freedom of thought can feel like a form of betrayal to the science you have been taught, with its laws and apparent certainties. But it has an essential role to play in the scientific process. In this context, the power of the hypothesis is that it provides a means of moving beyond both the comforts and the limitations of what we already know: a raft that may be rickety and is liable to be abandoned at some later point, but which nevertheless provides a vessel to row into the unknown and the as yet unconsidered.

When a hypothesis can begin with such a broad canvas as the entire spectrum of what is physically possible, or may fall accidentally out of experiments that were designed for a different purpose, you might be wondering where to start. How can scientists ever wrestle this mass of ifs and maybes into a coherent set of conjectures and a premise that can be tested? With the most complex problems, the answer tends to be slowly, in multiple stages, and through sometimes unexpected turns. The chain of ideas, observations and interpretations that aggregate into a workable hypothesis may be the work of many hands and many years. A discovery that begins in one place and may seem unremarkable can be the spur to something significant later down the line, when it is seen in a different context: a jigsaw piece from a different set that magically and unforeseeably falls into place.

As Katharina Schmack suggests: 'It's important to remember that science is chaotic, it's not linear. The weirdest discoveries lead to the strangest outcomes later on . . . At the time of discovery you will never know whether it will be useful or not.' She highlights the example of optogenetics, a pioneering field of neuroscience that entails the use of light to control individual cells in the brain. Described as a 'light switch for the brain', this amounts to a revolution in neuroscience and psychiatry, with the potential to deepen significantly and entirely alter our understanding of a whole host of conditions that originate in the brain, from mental illnesses to degenerative diseases and sources of neurodivergence. (Which I am all for us studying, as long as the intention is to improve understanding and not pathologize neurodivergence as a problem to be solved.) Although the method was not developed until the early years of the twenty-first century, it owed its discovery in part to conjectures that had been made a generation earlier, and discoveries that occurred a long way from animal nerve cells – a classic example of how a hypothesis may germinate very slowly.

The work done by Professor Karl Deisseroth to develop optogenetics in the early 2000s had its roots in developments from three decades previously. Moreover, these came in nothing like the order you would expect. As he has written, the idea came from one place and the early pieces of evidence from somewhere quite different. The essential hypothesis was that of the DNA pioneer Francis Crick, who in 1979 'suggested that the major challenge facing neuroscience was the need to control one type of cell in the brain while leaving others unaltered'. The then prevalent method of using electrodes to stimulate the brain was far too imprecise to achieve this. Crick toyed with the idea that light could serve this purpose, 'because it could be delivered in precisely timed pulses', but the idea remained unexplored, a kite that would not be grabbed until much later.[11]

The knowledge with which to do so, strangely enough, already existed. Back in 1971, researchers at the University of California, 'in a realm of biology as distant from the study of the mammalian brain as might seem possible', were studying the proteins produced by microorganisms (which include bacteria, algae and fungi). They were looking at microbes such as the bacteria from African soda lakes, and how they could survive in such harsh, saline environments. Their discovery – that one of the proteins acted as an effective regulator of electrical charges in response to certain forms of light – was a breakthrough whose wider relevance would not become clear for many years. It is ironic, as Deisseroth has written, that 'the solution to Crick's challenge – a potential strategy to dramatically advance brain research – was latent in the scientific literature even before he articulated the challenge'. Just as in life, where a word of encouragement or a comforting message can have much more impact than we realized at the time, the work that scientists do is often only obvious with the benefit of hindsight.

So it was with optogenetics. In the years that followed, more ospins (the light-sensitive proteins in question) were identified, with increasing efficiency at converting light into electricity. A research team in Germany first discovered an ospin in green algae, and then successfully introduced it into embryonic kidney cells, showing that it would respond to photonic stimulus.[12] Yet the distance between these inklings of possibility and an actual breakthrough was dauntingly large. There were enough questions and doubts that Deisseroth remembers other scientists considered trying to control the brain using ospins but disregarded it as too difficult. There were so many potential points of failure: the cells might not be able to adopt and produce the proteins as needed; the ospins might not work with the necessary efficiency and agility; there was no guarantee that an animal brain's complex machinery would respond in the same way as a single-celled organism. There was, he later recalled, an 'extremely high risk of failure'.[13] Yet having taken that leap, he found the early experiments with rat brains 'worked shockingly well'. In little more than a year, his team had published the first work outlining what would become optogenetics. From bacteria that live in very salty water, a breakthrough with the ability to transform our understanding of psychiatric diseases, overhaul our ability to control pain, and even equip doctors with new tools to restore sight, had begun to emerge.

The remarkable emergence of optogenetics is a wonderful example of science at its chaotic, brilliant best: a discovery in one generation helping to seed a breakthrough in another, via a conjecture whose answer was already in the process of being developed. It underlines the complex chains that are sometimes needed to tackle complicated problems: a testable hypothesis may rest upon a whole series of observations, discoveries and speculations made by different people in different places at different times. Work that may appear wasted, or an inquiry that appears to have run into a dead end, may actually turn out to be incredibly useful.

But the hypothesis itself cannot be so ephemeral. Even though it is far from fixed and may ultimately crumble under the weight of testing, the hypothesis needs a solid form that observation lacks. As the case of optogenetics shows, this can require the appetite to go where others will not: to propose a concept that seems outlandish, attempt an experiment that you expect to fail, or to risk criticism for having overreached. A hypothesis does not necessarily emerge from the existing evidence, falling off the production line of research by default. It may require someone (or some people) to manufacture it, pulling together the disparate threads of evidence and matching them to observable problems: scientists who are prepared to look silly or be proven wrong in the search for something potentially exciting.

A bit like being brave enough to ask someone out or express a differing viewpoint in a meeting, scientists need to have the courage to give voice to their wackier and more optimistic ideas. These might be entirely wrong, which is fine. They might be thought-provoking, helping others to develop their own thinking in a different direction or to challenge a wonky status quo. And they might *look* wrong in one context, or wildly unrealistic, but begin to make sense when another researcher with better information and tools can pick up the thread years later.

One lesson of the hypothesis is that we have to be willing to be bold and take risks. Our best ideas are worth nothing unless we say them out loud or write them down. Another is that the flip side of courage should be humility: the willingness to change course when the evidence suggests we may be wrong. Science shows us how hard it can be to pull a coherent thought out of the mass of things that surround us in life. Should we give up on a relationship that seems to be faltering, or just accept it's a bad patch to get through; does a friend need our honest advice, or does it risk making things worse? So often we are faced with uncertainty and conflicting evidence. The scientific approach is not to rush to conclusions but to gather the evidence you have and develop a hypothesis: a rope you can hold on to, but not something that limits you from changing your mind later. A guiding light but not a complete roadmap. And with that, you have officially reached the starting line. Congratulations. Now we just have to put it to the test.

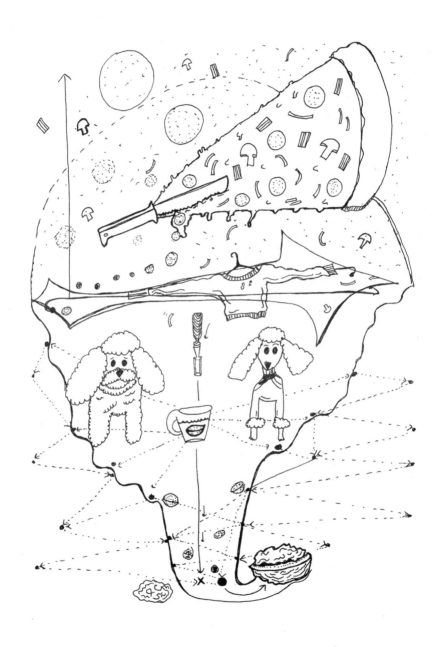

3. Focus

How to approach problems, handle data and pick a path

Having an ADHD brain makes focus one of the most elusive things in my life: not the biggest help when your job consists of sifting through data in an orderly way, searching for patterns and hidden signals among the noise. For that, it helps to be able to concentrate. But usually my mind is hopping all over the place, trying to pin down a hopelessly broad concept one minute and getting lost in an irrelevant detail the next. I pogo from macro to micro and often forget to stop in between (sometimes I can't remember why I am doing it in the first place).

Getting my mind to be still, so I can see clearly, is one of my biggest challenges. But it is also fundamental to how I think and work as a scientist. I want to be able to pile new dimensions and ideas onto each other like toppings on a pizza, to wander round inside huge problems and to study small ones until the complex systems contained within them become clear. A small, specific and limited project always feels like the enemy: an itchy jumper that has shrunk in the wash, making it hard to move your arms.

This expansive approach doesn't always work. Sometimes you cast your net so wide that you lose sight of what you are trying to drag in. You might get called impatient, accused of going off on a tangent, or quietly reminded that you still have a deadline. It can feel like the science equivalent of telling a friend you're leaving the house just as you are about to spend half an hour deciding what to wear. But for me, and many others, this is what works. You have to make the problem you are trying to solve as big as possible, looking behind every cushion and under every pile of papers, before you can decide where you should start. Work that can seem aimless is often leading you to exactly where you need to be.

Whenever I am doing this, I have to remind myself that all this inner rattling is a thought just trying to come out of its shell, and that you often have to stretch out a problem before you can contract it down to something manageable. Like this book, where I needed to know what would be in the last chapter before I could write the first.

Luckily I'm not alone in this. Anyone doing the legwork of science – the experiment design, the evidence gathering, the proof building – must grapple with focus, the question of which approach to take, what information to prioritize and how to go about it. Essentially how to dig into your hypothesis: making a scientific problem or inquiry both big enough to be interesting, and manageable enough that you can do something practical with it. In doing this, there are specific choices to make but also an overarching dilemma to address: Do you run straight towards the problem you have set yourself (as best you understand it), or do you try and walk around it, seeing it from every possible angle? Is it better to quickly narrow your perspective for the sake of clarity, or more beneficial to keep your options open and avoid closing off paths that may subsequently prove useful?

In the vivid metaphor of the French mathematician Alexander Grothendieck, a problem to be solved is like a nut that you need to open, 'the nourishing flesh protected by the shell'. He suggested two ways of going about this. You can try to break it open with a hammer and chisel, hitting it in multiple places until it finally cracks. Or you can simply immerse the thing in liquid and let time do its thing. 'The shell becomes more flexible through weeks and months – when the time is ripe, hand pressure is enough, and the shell opens like a perfectly ripened avocado!'[1] (You can think of him as the original smashed-avo-on-toast guy.)

Grothendieck's point was that the direct approach isn't necessarily the best one. Simply trying to apply the most obvious solution to a particular problem may give you the *feeling* of making progress (after all, you're holding a chisel) but it isn't necessarily getting you anywhere. By contrast, the roundabout approach that appears unpromising, which is patient and above all indirect, may ultimately be the winner. Whereas the chisel holder has made the problem their entire point of focus, the slow soaker has zoomed out quite a lot. They work on other things surrounding the problem, which may or may not affect it. They take a broader view, and seek a bigger prize. By the time they have earned it – developing their theory or method – cracking the nut they started with has become almost incidental. It just happens. As Grothendieck wrote, this approach is a little like the rising tide: progress is made gradually and then all at once. '[The] sea advances insensibly in silence, nothing seems to happen, nothing moves, the water is so far off you hardly hear it . . . yet it finally surrounds the resistant substance.'

It is not just his memorable turn of phrase that makes Grothendieck someone worth listening to. As well as being handy with a metaphor, he was one of the most influential mathematicians of the twentieth century. His eccentric lifestyle – living off austere diets that sometimes consisted only of dandelion soup, and at one point limiting himself to speaking only in rhymes – did not prevent him from making groundbreaking contributions to his field (and possibly it helped). A reminder not to self-censor for the sake of appearances!

His work in algebraic geometry – visualizing and exploring algebraic equations as circular shapes – has been widely recognized as groundbreaking, paving the way for subsequent advances including the resolution of one of the fundamental theories used in mathematics and public-key cryptography, Fermat's Last Theorem. 'Whole fields of mathematics speak the language that he set up,' the Stanford Professor Ravi Vakil has said of him. 'We live in this big structure that he built.'[2]

Much of that success has been credited to his willingness to eschew the hammer for the long, slow soak: not limiting himself to the tools that others had provided, but instead fashioning his own. (He also wrote about being the kind of mathematician who was much happier building houses than redecorating those that others had constructed before him.) Grothendieck, his supporters have argued, was the expert at finding the sweet spot between so general as to be meaningless, and so specific as to be useless in any other context. 'He just had an *instinct* for the right degree of generality,' the US mathematician John Tate suggested. 'Some people make things too general, and they're not of any use. But he just had an instinct to put whatever theory he thought about in the most general setting that was still useful.'[3]

His work, and his philosophy about how to go about it, lead us to an important corner of the scientific process: the question of where and how to focus an inquiry. His avo analogy is clear about the dangers of being too single-minded, insisting that there is only one problem and likely just a single way to solve it. Yet the opposite is also true, where generalization becomes abstraction to an unhelpful degree, and your canvas has grown so broad that no amount of paint will fill it up. Scientists are looking for the holy grail that is general enough to be widely applicable and detailed enough to be meaningful. But how do they strike this balance? And what can we learn from their efforts to draw the biggest possible circle while still ensuring that the ends will, eventually, meet?

The question of how to focus a research effort begins and ends with the question of data: what can you access or generate, how relevant is it, and what are you going to do with it? In clinical trials, researchers create their own data, experiment by experiment, doing something that has never been done before and forming an entirely novel data set as they go. Yet much of science faces the opposite problem: not a lack of available data but a surplus, a pool so deep and wide that any unwary researcher could quickly become submerged and lose all sense of direction (sometimes while still believing they don't have enough data). It is not just the quantity of data that causes confusion, but something data scientists call dimensionality. When you are exploring, for example, demographic data sets covering an entire population, you quickly run into numerous factors (dimensions) that can't all be plotted together on a simple X and Y axis. There are age, ethnicity, domicile, education, health, income, family size and tax status to name just a few. In other words, a great tangled mess of information which may contain a wealth of insights, but only if you can first wrestle it into a more manageable form. The temptation may be to go off and crash your computer by opening a spreadsheet with over 150 features (guilty), but the reality is that you can never capture everything you might want in a single data set. There is always a requirement to sort the data somehow, using a combination of instinct and subject knowledge to identify the most important variables and bring out the information that matters. A little like cutting a diamond to bring out its brilliant shine, the way you slice and dice a data set is fundamental to how much insight you can derive from it.

The technical term used to describe this work – data wrangling – is nicely descriptive of the task. You are wrestling an octopus of information and, before you can begin to analyse it, first have to get it into some kind of accessible format: combining sources, discarding what appears to be irrelevant and structuring what is left into something that analysts can comb through and use. Even at this early stage, you are making decisions about where to focus, what to ignore and how to organize the information at your disposal. You are at risk of omitting something that may be important, or formatting the data in such a way that will skew subsequent analysis. It's a bit how I imagine a sculptor must feel, knowing they need to start chipping away at the solid block of material, but worried about cutting away an important bit that can't then be added back.

That pitfall only becomes more apparent as you move from initial organization to simplification. Even the initial wrangling may still leave you with lots of data features to choose from. One method of further clarifying the material is a data science technique known as dimensionality reduction, which does as the name suggests, eliminating dimensions until the data has become something that you can see clearly and begin to work with. This not only helps to make the data easier to visualize by grouping it together into useful clusters, but it also removes bits that are likely irrelevant for the purposes of your work. As Dr Leland McInnes, who helped create a dimensionality reduction technique called UMAP, has put it: 'We get to go from something complicated . . . to something simple and combinatorial that we can work with easily, and yet we retain all the information we want to care about.'[4] In very simple terms, this means that graph data that may begin as a massive mess of dots plotted everywhere – a bit like a Jackson Pollock painting – can be turned into something represented as relatively neat groups of clusters.

UMAP functions by using machine learning to trawl a high-dimensional data set and replicating it in a low-dimensional format (i.e. something less noisy and with fewer variables, in which useful data points are grouped together). Its goal is to hack away at the thickets of data to clear a visible path, at the same time as preserving important areas of connection – which it does by weighting the probability that different data points are linked, and that those linkages may be important.

Of course, this kind of aggressive filtering can come at a cost. You are simplifying, which is necessary and feels good, but must also remain wary of losing something important: closing down a pathway that might have led somewhere interesting. The moment you have reached for what Grothendieck called the chisel, you become aware that you might be chopping off something that you will later discover you need. 'No matter how you frame a problem, there's always a different lens or way of looking at it. And every lens makes some things clear and obscures others,' is how Dr McInnes put it to me.[5] Of course, UMAP – which played a role in helping to isolate and identify variants of SARS-Cov-2 (the Covid-19 virus) – has been designed to embrace and rationalize this dilemma.[6] 'As researchers at Google have summarized: 'To determine connectedness, UMAP extends a radius outwards from each point, connecting points where those radii overlap. Choosing this radius is critical – too small a choice will lead to small, isolated clusters, while too large a choice will connect everything together.'[7]

So we can see that dimensionality reduction is a little like gardening. To ensure your flowering shrub will bloom in season, you need to give it a good hard prune at the right time (mind your fingers), but without cutting away so much that there is not enough left to grow. This is one of the perennial balancing acts of any scientific process. You have to narrow down the parameters of your inquiry – to give it a focus – otherwise you will never see anything. But if you narrow down too much or too quickly, you will limit your ability to discover anything worth knowing. This is the problem framed by the mathematical theory known as optimal stopping: trying to calculate the correct moment to do something such that you maximize your chances of achieving the intended outcome, and limit the possibility of everything going wrong.

It's also one of the great challenges for an ASD mind like

mine, which naturally wants to collect as much data as possible and get lost in it and mix it up in different combinations. I want to find out everything I can and, in the process, constantly worry that I might be missing some important variable – kind of like an analytical FOMO. When I'm asked to summarize and simplify what I've found, a serious case of brain freeze can follow. This was a particular battle while writing my PhD thesis, when I kept trying to cram in more information, until my supervisor pulled me out of the weeds with a pertinent observation: I was compensating for a lack of clarity with a surfeit of detail. She even came up with an analogy I could use: the thesis was like a dog that had gone so long without a trim that you could hardly tell what breed it was. Sticking on more hair wasn't going to solve anything; instead I needed to 'trim the poodle'.

I'm not the only scientist who has struggled to get the balance right in their work. While we may assume that scientists move from A to B in organized fashion, twiddling the dials on their microscopes and zooming in unerringly on the answer, the reality is a lot messier. Often it entails going around in circles, casting about for clues, not entirely sure of your direction and constantly doubting yourself. Just because you have developed a hypothesis doesn't mean you will know where to focus your inquiry and on what. There is still a lot of trial and error involved to get to the starting line proper of a research project. As Leland McInnes told me, he believes the early stages of an inquiry can still involve a lot of wandering around: 'You feel like you're trying to climb a mountain to get to the top, and you climb up a way and then suddenly realize you're at a big plateau. And no matter what you do, no matter what direction you head, you make no progress whatsoever. So you just have to keep trying. Because even though every direction seems to be flat and [offering] no improvement, eventually you do one thing, and it takes you up to the next stage, and all of a sudden you're climbing again. But it's not clear at the outset what that thing is going to be.'[8]

This need for patience was a regular theme in my conversations with scientists about their work. Several spoke about how a published paper had come about in several stages, with an idea only coming to fruition years after the observation that had first prompted it, when new data became available. It is not always possible to hammer away at a problem and hope to crack the nut there and then: to pick one path and expect that you will be able to stick to it without deviation. The time may not be right, the methods or tools may not be sufficiently developed, and your own knowledge may not yet have matured. Trying to find the right focus and direction for a new research project is a bit like making life choices in your twenties: you can be overwhelmed by the number of different paths branching ahead of you, and feel like whatever you choose will set your destiny for ever – the beloved quarter-life crisis. But in life, as in science, it's never as simple or final as that.

The scientific process is invariably non-linear and can be long and drawn out, with hypotheses sitting on the shelf until the time has come to dust them down, if that ever finally arrives. Sometimes you sit waiting for the tide to come in, but it never actually does. For every new theory and paper, there are many that fell by the wayside for lack of time, funding or data. Like the music industry, the hits are few and far between and don't always come when or from where you expect. It's important not to be discouraged by this: the fact that not every idea comes to fruition isn't a good reason to have fewer of them. Scientific breakthroughs ultimately depend on people working away without the immediate reward of achieving anything tangible: trying things that don't work (but might still be useful) and devising bits of a solution that won't be relevant until some undefined point in the future. Such patience is the bedrock on which progress is eventually achieved – stitched together from all those loose bits of fabric that could so easily have been discarded.

Next to patience comes the need for ignorance, or at least to accept the reality of it. In his widely read paper 'The Importance of Stupidity in Scientific Research', the microbiologist Martin Schwartz wrote about how his PhD taught him the extent of what scientists do not know or understand, even when they are leading experts in their field. His light-bulb moment came when he consulted on a research problem with a senior colleague, who would subsequently win the Nobel Prize and 'knew about 1000 times more than I did'. When the expert said that he had no answer for him, Schwartz started to realize that there might actually be an upside to *not* knowing. 'If our ignorance is infinite, the only possible course of action is to muddle through as best we can.' His message is both simple and compelling. If you're not making yourself feel stupid in the course of scientific research, you're almost certainly not doing it right, not asking the really hard questions or going far enough beyond what others have already done. By contrast, as you push into the realm of 'absolute stupidity' – into areas of inquiry where no recognized answers or solutions exist – you give yourself the best chance of unlocking your potential as a scientist. 'Productive stupidity means being ignorant by choice . . . The more comfortable we become with being stupid, the deeper we will wade into the unknown and the more likely we are to make big discoveries.'[9] As this suggests, any scientist worth the name has to get comfortable living in a kind of limbo, building in the space of what is unknown and potentially impossible. It's wonderful, frustrating and – when it works – completely amazing: a reminder that science is as much an emotional pursuit as a rational one, in which the researcher must manage the temperature of their own mood as much as they wrangle cold, hard data.

(To indulge in detail is like taking the first few scoops of ice cream – best done in moderation, resisting the urge to consume the whole tub.)

The importance of both patience and ignorance shows how there is something almost organic about a scientific inquiry, an iterative and slow-moving quality that defies the neat boundaries of hypothesis, experiment and conclusion. Even when you are trying to focus your research, you may have to accept that things are going to be fuzzy for a while. You may have a good hypothesis in your pocket, but you have to keep hypothesizing about how best to test it. There is no smooth progression in scientific research from uncertainty to clarity. The scientist must continue to work and experiment, never quite knowing if they are knocking on the right door, or what variation on a theme will make the music they have been looking for. Most scientists would admit that a decent proportion of their work consists of stabbing in the dark, taking educated guesses and just seeing what would happen if you tried *that*. These might be very minor experiments, Leland McInnes suggests. 'I admit I spend time just fiddling with things and seeing what works . . . Rather than trying to come up with concrete theoretical things to do, you just do all the things in software you can, add knobs and tweak things, do whatever you want to make it better, and then eventually you get a little bit better and then you go back. And you alternate between this engineering hill climb when you just build things, and then back to a theoretical thing. What's the theoretical explanation that explains *why* this would ever work?' This process of trial and error in the experimentation, and then retrofitting the theory as you go to illuminate what you have done or discovered, illustrates how science is a circular rather than linear pursuit. There is no question of moving in seamless stages from observation to thought, hypothesis, experiment and result. Instead you are hopping like a rabbit back and forth, updating priors and tweaking assumptions, the picture getting a bit clearer with each circle from theory to building and back that you complete. You can't fall in love with the first draft of the story you are trying to tell or the way it looks from the outside, because the nature of the research process means that something you discover at the end could easily transform the way you write the beginning.

This is the strange, crab-like dance that is science: scuttling round and round in circles until you see something worth exploring further, at the same time as trying to move towards a result or outcome of some sort. You are both going forwards and going round the houses all at once, often unsure which way is up or how long it is going to take. You can't just zoom in on something, you also have to zoom out and work out where all your efforts have taken you. Yet for all the importance of patience, letting ideas mature and answers gradually rise to the surface (like frozen peas is boiling water), there is ultimately a need for decisiveness: to start wielding the chisel, chopping away at the block of data and trying to sculpt some representation of your initial hypothesis from what is left. (And if individuals aren't driven to do this of their own accord, the limits of their funding and the requirement to publish provide a pretty firm shove.) It's at this point that things get interesting, but also risky. The narrowing down of data and methods does not just carry an opportunity cost of potentially better ideas left unexplored; it can also lead to outright bad science, developing algorithms or deducing conclusions based on data sets that cannot be replicated in any other meaningful context. That makes the way we choose to focus not just one of the most important parts of the scientific process, but also potentially one of the most dangerous. Next we'll look at how data scientists go about this, and the things they work hard to avoid: conclusions that don't match up closely enough to the available data, and ones that are problematic for the opposite reason, because they fit too well.

The question of why we dream and for what purpose is one that has animated neuroscientists (as well as artists, novelists and pillow makers) for generations. The mystery of what the brain is doing while we sleep has produced a slew of hypotheses: that dreams are the way we regulate our strongest emotions, a safe place for us to play out frightening thoughts; that they are a way for the brain to sift and store its memories, consolidating what we have learned by replaying it in our heads; or conversely that dreams are actually a vehicle for 'reverse learning', and a way of forgetting things we would rather not remember (like half the things we did in our twenties). Some have argued that, as a form of mental simulation, dreams act as one of the brain's problem-solving mechanisms, and others that they are a way of the brain trying to predict future states. The neuroscientist Erik Hoel has put forward another explanation. His 'overfitted brain hypothesis' (OBH) argues that the brain is not trying to achieve something by dreaming so much as to avoid something: the fate of 'overfitting', a term from data science describing an algorithm that has been trained on such a narrow set of data that it cannot apply what it has learned in any other context.[10] Because of this, it will perform considerably worse when presented with unfamiliar data – like learning the piano in such a way that you will only ever be able to play 'Chopsticks'.

This is a familiar dilemma in artificial intelligence: the 'trade-off between generalization and memorization', as Hoel describes it. In simple terms, the more you optimize an algorithm to perform within a particular data set, the less capable it may become of transferring those capabilities in a different context. The better you fine-tune an AI in its 'training' environment (where it learns, for instance, to tell the difference between another car and a pedestrian), the more useless it may become when applied to any other task or set of information. More focused may also mean more limited. Hoel's contention is that the methods AI researchers have developed to overcome this problem – 'the injection of noise and the corruption of input during learning' – are what the human brain has been doing for itself all along. Through dreams, he argues, the brain is essentially trying to escape the limitations of what it has experienced that day (its training data), and to avoid the overfitting that could lead us to struggle when placed in a similar but different situation.

This would explain both the similarity and the strangeness of dreams: how they often appear to relate to the experiences of the previous day, but at one remove from reality and with bizarre twists in the tale. All the odd bits about dreams can be linked back to the brain's effort to avoid overfitting: their 'hallucinatory nature' (deliberately altering what we actually experienced), their 'sparseness' (losing some bits of the story as it actually happened) and their 'narrative nature', because the brain 'understands reality in the form of events and stories'. For example, if we have been out in the car that day, we are more likely to have a dream that somehow involves driving, but in a markedly different context – avoiding the overfitting of our familiar routine and helping to generalize the experience. As Hoel suggests: 'It is the very strangeness of dreams in their divergence from waking experience that gives them their biological function.'[11] Through dreaming, our brains are not just trying to entertain or confuse us; according to this hypothesis, they are trying to save us from ourselves, making us robust creatures that can adapt to different conditions rather than being siloed by the way we learned something in reality. And if my boss is reading this, that's definitely the reason I sometimes take cat naps in the office – what my friend, who likes to have a coffee first, calls the Nappuccino.

Hoel's hypothesis for how and why we dream helps to shed light on the scientific problem of how to process information and focus an inquiry. In any research work, scientists are constantly in danger of either under- or overfitting: grabbing straws in the wind that turn out to be simple correlations, or fitting conclusions so closely to a certain set of data that they lose their integrity and wider applicability. In the latter case, that can lead to findings that tell us about nothing except how a researcher chose to pare down a particular data set. When data is being filtered and its dimensions reduced for the sake of making it easier to work with, one group of neuroscientists has written, 'there is the danger that the choice of a cut . . . may create apparent signal where none actually exists'.[12] Scientists are forever in danger of being damned by too little data, or condemned by the approach they took to handling too much. Focus too little and you will see nothing useful at all; but focus too much and too narrowly, and you might either miss the point or exaggerate something that isn't particularly important.

Data scientists have developed a number of approaches to try and avoid this fate. (Not for nothing did the *Harvard Business Review* call it the sexiest job of the twenty-first century.) There is the 'lock box' of data carved from the initial set that will not be touched until the algorithm has been fully developed and trained using the core data, at which point it will be opened for testing to determine if overfitting has occurred. Another approach is 'blind analysis', in which some aspects of a data set are either obscured or interfered with – altering labels, removing some parameters or somehow adding 'noise' into the system.[13] In both cases, the principle is to actively try and *avoid* the locally perfect outcome, for fear that once it has been achieved, it will become set in stone and not allow the AI to adapt in any meaningful way. (A procrastinator's charter: lovely!) Scrambling the signal a little, the logic goes, increases the chance that it will also be able to tune into different frequencies rather than getting stuck for ever on the same channel. A little imperfection makes for a more resilient and adaptable model. Difference and jittering noise support this adaptability – one reason why I never apologize for my neurodiversity, or for being that *little* bit too random. (Invariably, the small minority of people who have a problem with it are the ones who would benefit most from its influence.)

In that there is maybe a wider lesson about science and the research process. We never know how the work we are doing, and the detail we have chosen to focus on, may be used by someone else in the future. A theory that was developed by someone for one purpose could easily be picked up by another researcher years later and applied to something else entirely. Neuroscientist Rogier Kievit gave me the example of Brownian motion, a theory devised to understand the movement of particles suspended in a fluid (such as dust mites in the air) and how this has been cross-fertilized into his field, helping to illuminate the way the human mind accumulates information and makes decisions.[14] Drift-diffusion models, which are used to estimate how people will make decisions and how quickly based on the information they encounter, are based on the same principles as Brownian motion, which shows that the random collisions between molecules will on average lead them to disperse evenly through the fluid. In the drift-diffusion model, the equivalent of dust particles colliding with gas molecules in the air is how our brain encounters new information and stimulus. All these 'collisions' move us towards either an upper or lower boundary that signifies the poles of a binary decision. The 'drift rate' – how quickly we accumulate information and therefore move towards one of those decisions – is determined by a combination of how easy the information is to access, and how accurate, credible or familiar it is.[15] In this way, a theory that has its origin in the work of a Scottish botanist working in the 1820s remains fundamental to how the neuroscientists of the twenty-first century understand the decision-making processes of the human brain.

It's a beautiful reminder that no scientific work ever stands alone. Every hypothesis, finding and theory is part of a continuum that stretches back as long as people have striven to explore and understand the world: an open-ended process in which principles are questioned, theories are reshaped for different purposes, and ideas may be stitched together from disparate fields. It's like a book that is continually being written and rewritten by each generation of scientists: old pages ripped out, new ones added, sections rewritten and notes added in the margins. That is why scientists work so hard to avoid the curse of overfitting, and developing work that has no wider relevance or applicability. A scientist is not really trying to develop the perfect sculpture that will stand on its own, but more of a tool or puzzle piece that other people can pick up and use in their own ways – including ones they had never foreseen. You might be focusing in on an idea, a problem or a data set with a high level of specificity, but if your work becomes so granular that it doesn't apply to anything else, it's like tuning in to a radio frequency that no one else can access or wants to listen to (unlike my guilty pleasure of Magic FM). In this way, science can be seen not just as a search for answers but as a quest for relevance: to discover and verify something that, however incrementally, contributes to the wider development of a field of study and a body of theoretical work. In the overall context of their field, the work done by most scientists in their career will amount only to a few small particle collisions in a river which is full of them, all happening upstream and downstream, in and out of sight. The ability to sit on that riverbank, dangling your feet over the edge of all this bubbling wonder and frustration, is what makes it all worthwhile – *just*.

The search for signal, therefore, is one of the most brain-scratching and contradictory parts of the scientist's job. You need to chop down some trees and clear a path, in the constant knowledge that you might be destroying something important. You need to hone a conclusion specific enough to be notable, without being so obscure that it's no use to anyone else. And you need to avoid the trap of delivering a finding that doesn't tell the wider field anything except how the data was sifted and sorted in that particular instance. On top of that, as Leland McInnes suggested, often researchers will not feel in a position to do anything at all: stuck on their mountain plateau, looking at data that appears to hold no insights, and wondering whether to press on or give up. The fear that you might be doing something wrong competes with the worry that you might be getting nowhere at all.

We could all learn something from how scientists grapple with the dilemma of how to make their work accurate, relevant and interesting at the same time. Because if you think about it, our entire lives could be described as a process of gathering and filtering data: meals we like to eat, people we like to spend time with, hobbies we enjoy and the career we want to pursue. If you're the kind of person who compiles lists of pros and cons when making a big life decision, then that is nothing less than an exercise in data collection, dimensionality reduction and sorting – clustering together all the factors that are pushing you in one direction (remember drift diffusion?) and those that are pulling you in another. It's ironic that I do this for a living, since outside the lab it's the one thing I can't do to save my life. The neurotypical ability to deal in aggregates and abstractions is something that many neurodivergents (I'm putting both hands up here) struggle with. I'm generally so busy adding new dimensions to the data set about what midweek meals I like to cook, or my dog's favourite walking routes, that the idea of boiling it all down to spit out a single conclusion can be terrifying, and I have to give myself another talking-to, so I can remember that you don't have to reinvent the world every day, and there is a life beyond all these extra dimensions. Sorry, Wendy, for those days when we end up doing a fifth walk just to make sure.

But I don't think I'm alone in this struggle, even though my ASD accentuates it. To some extent we all grapple with this problem. And as we sift through the data of our lives to try and plot a path through it (with our brains working in the background to generalize those experiences), we might take some comfort, as well as a few pointers, from the researchers who spend every working day doing the same.

The lesson is to be aware of under- or overfitting: grabbing too quickly on to data points that couldn't possibly signify a trend (like meeting someone for the first time and deciding you will be friends for ever), or completing a perfect model in a data set that will never be repeated, like the music you loved listening to when you were fifteen, the drink that tasted amazing on holiday, or the summer fling that felt magical but which just wouldn't have worked in any other context. Much as scientists are constantly aware of drawing the wrong conclusions from their data, we have to be honest with ourselves: is this really what the evidence leads us to believe, or have we failed to take context into account (or simply drifted into wishful thinking)? Would you let someone else repeat the experiment in the expectation that they would come to the same result? (Would you have the courage to tell your mum or best friend what you had decided, and why?)

Science can help us think about how to make decisions and untie the knots of difficult problems. And it can offer a comforting arm around the shoulder as we struggle with these, reminding us that it's OK not to be sure, and that the perfect answer is usually more of an illusion than a reality. Some of the smartest people in the world spend whole days scratching their heads, fiddling around with models and hoping that something new or interesting will emerge from that work and the hours spent staring at the same wall. They don't know what, how or when the beginnings of an answer will become clear; they just recognize that the only thing worse than not knowing is not doing anything about it. Just like life, science doesn't move at a steady pace or in a single direction. The only way to make progress is to keep coming back, trying new things, and recording how you did. And on a good day, data that seemed impossibly noisy starts to give the faintest hum of a signal. Just the right twist of the microscope makes a blurry image a little clearer. You begin to gain the precious asset of focus, and the clarity it provides about what to do next.

4. Interpret

How to make sense of information

Sometimes in life we get it wrong. Whether it's waving at a stranger on the street you thought you recognized, believing that an acquaintance who says we 'must do lunch' actually wants you to book a restaurant, or misreading the 'signals' from someone who you thought was interested in you, we have all made heroic leaps of misinterpretation, grabbed the wrong end of the stick and ended up looking pretty silly. In science it's no different, except the red faces usually come not after a momentary lapse but after perhaps months or years of research, multiple published papers and endless debate about the implications of what appears to have been discovered.

However awkward or embarrassing, these missteps are a part of scientific research, especially when challenging old narratives and attempting to break new ground. Even apparent experts in decoding data and interpreting evidence can completely misread the information in front of them. Sometimes those mistakes will be minor and obscure, in areas that are only going to be seen by subject-matter specialists. But others are not so lucky, and happen under the full glare of public and media attention. In the 1960s, a group of scientists fell into the second category, swinging and missing in a pretty remarkable way. Go with me on this one: they tried to reinvent water (and, for a time, believed they actually had done). This is the story of a now-forgotten substance that at the time was believed to be revolutionary. It was called polywater.

This strange episode followed a scattering of experiments in which researchers had noticed that, when they isolated water in miniature vessels, it did not evaporate as easily as it should have done. In 1961, the Soviet scientist Nikolai Fedyakin appeared to verify this observation with an experiment in which he first evaporated then condensed water into capillary tubes the width of a human hair. What was left at the end was not water as we know it, but something that appeared denser: 'as gooey as petroleum jelly and about one-and-a-half times as heavy as the stuff that comes out of your kitchen tap', as an article in the journal *Popular Science* later described it.[1] This 'offspring water', as Fedyakin initially named it, gradually piqued the interests of scientists first in the Soviet Union and then around the world. It not only looked unlike water but also behaved differently – it had a much higher boiling point and lower freezing point, and did not expand when frozen as H_2O does. Scientists in the UK and USA were able to replicate its formation, and in 1969 spectroscopy experiments – which measure how matter absorbs light and radiation – appeared to confirm that it was in fact a novel substance.

It was at this point, almost a decade after Fedyakin's initial work, that the mysterious molecule briefly became a phenomenon. When the two American scientists behind the spectroscopy experiment published their paper, they coined the term that would stick to this viscous substance: 'The properties are no longer anomalous, but rather, those of a newly found substance – polymeric water or polywater.'[2] This combination of fresh evidence and a neat shorthand helped to set off a rollercoaster of activity and debate. The US government handed out a grant for polywater to be manufactured at scale (it had only ever been produced in minuscule quantities), scientists hypothesized about the chemical structure that might explain its properties, and both industry and the media began to speculate about it.

Polywater might be present on the moon and it could be a potentially dangerous contaminant if it got into the water supply (one physicist described it as 'the most dangerous material on earth').[3] It might even hold the secret to reversing the ageing process.

Like many good stories, none of these predictions turned out to be true. Both the optimists and the doom-mongers were blown out of the water by a team at the University of Southern California, who conducted new spectroscopy experiments on polywater samples they had produced. The result was unambiguous: put under a laser, this 'water' burned and turned black. It could not be a substance made simply of hydrogen and oxygen molecules arranged and bonded in some new way. Further testing revealed that polywater in fact contained impurities including sodium, chlorine and sulphate. The argument was ended when one of the scientists involved, Denis Rousseau, performed a simple and winningly definitive experiment. After a game of handball, he squeezed the sweat from his T-shirt into a flask and put it in a spectrometer. The results from the sweat off his back were effectively identical to those that had been found for the polywater that scientists had been so painstakingly producing under laboratory conditions. 'The implication was obvious: that the contamination of polywater resulted from the condensation of bio-organic matter on the surface of the freshly drawn capillary tubes,' he later wrote.[4] In other words, tiny amounts of perspiration had got into the water that researchers had been evaporating and condensing into the miniature tubes to make polywater. Scientists who thought they might be leading a revolution had 'been fooled by the sweat that dripped from their own brows'.[5] Almost in an instant, a potential scientific miracle, one that had been brewing for the best part of a decade, collapsed into nothing. A helpful reminder that science is indeed achieved through blood, sweat and tears: in this case quite literally the second, and almost certainly the third.

The story of polywater is not just a quirk of science history, but a parable for the scientific process and how it can go wrong, even with lots of brilliant minds engaged with the best of intentions. It shows how the narrative doesn't necessarily follow the evidence, but can too easily be used to justify the experiments scientists do and warp the results they produce. It is, suggested the man who debunked it, an example of 'the loss of objectivity that can accompany the quest for great new discoveries', and an example of 'pathological science' in which researchers allow bias, wishful thinking and their desire to be pioneering to override some fundamentals of accepted scientific methodology.[6] When engaged in this, Rousseau suggested, scientists may be so swayed by their desired narrative that they refuse to conduct 'definitive experiments' that could prove or disprove a thesis, such as his sweat test. 'The investigator never finds the time to complete the critical measurement that could bring down the whole house of cards.'[7]

Polywater is a reminder that science ultimately comes down not to the collection of data but the interpretation of it. The history of research is littered with anomalies that could have been significant but turned out to be meaningless, quirks in the data that needed to be explored but ultimately led nowhere, and potentially dramatic discoveries that ended as damp squibs. Scientists should be the objective observers of all this, calmly processing information, thinking sceptically and stoically, and requiring minimum thresholds of evidence before proposing conclusions. Yet researchers are also human: they like to do interesting work, they are intrigued by new possibilities and they want to change the world. However professional and well trained they may be, scientists are susceptible to the same traps of peer pressure and wishful thinking as anyone else. They want to pursue a fashionable line of research for the same reason posh supermarkets will sell out of an ingredient days after Nigella has endorsed it. The herd instinct applies as much in science as it does with blockbuster films: when something is hot, everyone wants to be the first to see it. In rare cases like polywater, that can lead to lots of very smart people chasing limited evidence down the same rabbit hole, because it promised the answer they were hoping to prove. Unconsciously, they misinterpret the evidence until it says what they wanted to hear.

The importance of objective interpretation also manifests itself in more subtle ways: any scientific experiment begs scrutiny of how it has been designed, within what constraints, and how the results might have been changed by alternative methods or different parameters. Scientists must constantly and rigorously interpret the data they generate and the findings they produce, prodding and poking them for potential flaws or oversights. Equally, those in more experimental and theoretical fields sometimes need to make creative interpretations based on limited data, drawing extrapolations and assessing implications – not so much scrutinizing what they can see as trying to look around corners and estimate what they cannot.

Misinterpreting evidence in favour of an ambitious hypothesis is not the researcher's only pitfall. At the other end of the same spectrum is failing to believe data that is telling you something extraordinary, and which suggests that a cherished scientific consensus may in fact be false. Disregarding data that could be consensus busting because it 'feels' wrong is as big a trap as fitting limited evidence into a preconceived pattern. Scientists can stumble both when they trust too much in their data, and when they have too little faith in it. And that's before we get to the question of whose voice gets heard and which opinions are taken most seriously. Science is no different from any other line of work: there is a tendency for established views and establishment people to shape the narrative. When you try to diverge from what is accepted, some people are always quick to call you emotional or unrealistic, even though it's only by seeing past the consensus that we ever make progress.

For scientists interpretation is both a blessing and a curse, something that can cause problems both of commission and omission. A scientist cannot (or should not!) be so bold or broadbrush that they shape the evidence in front of them into a false premise, but nor can they be so timid that they fail to connect the dots in creative and sometimes unusual ways, to open up new ideas and ways of thinking. Over-caution will get you nowhere, making you a victim of your own limiting beliefs. But over-eagerness may take you to a place that the evidence doesn't ultimately support. The paradox of interpreting evidence in science is that it is both the most important part of the job, and the one that puts you at greatest risk of making mistakes. Great scientific breakthroughs are forged from the same desire to break the mould and believe the improbable as embarrassing failures like polywater. A single-minded belief in a new idea can be what's needed to fuel discovery, but it can also act as a brick wall you will bash your head against. The line separating a creative and brilliant interpretation of data from a nonsensical one can be incredibly thin. Nobody said it was easy!

If you've ever sat through one of the abstract-reasoning tests where you have to say if the next shape in a series is the green cube, red dot or blue banana, you might have wondered what exactly is being measured (along with when is it time to go). Can your suitability for a certain type of job be properly assessed by something so apparently random and unconnected to the real world? These awkward puzzles don't just tell us something about our intelligence. They also shed light on the layers of interpretation that scientists need to undertake in an experiment – working out not just what the data appears to show, but the reasons for it, and whether those are liable to change under different conditions. No scientist worth their salt simply takes a set of results at face value: they prod, ask how robust the numbers are and what they depend on. They wonder what would happen if they introduced this variable or removed that dependency. They interpret proactively and with scepticism, looking for reasons that data might not hold up to scrutiny, as well as searching for what it might ultimately tell us.

The good news (or bad, if you hate puzzles) is that there is decent evidence for such tests showing us something useful. In the late 1990s, researchers in Scotland tracked down survivors from a remarkable experiment. In 1932 and 1947, almost every eleven-year-old in the country had sat an IQ test; now they were asked to do it again (the tests were sat several times, at intervals a few years apart, with the original subjects now in their seventies and eighties). Not only were the results mostly consistent across this time lapse (demonstrating a correlation of between 0.6 and 0.7, where 0 is no correlation and 1 is total), but they were also indicative of how the children's lives had developed in adulthood. Those who had done better on the tests were notably less likely to experience early mortality: girls whose IQ scores had been two standard deviations (30 points) higher at eleven proved twice as likely to still be alive aged seventy-six.[8] Across the board, children who had scored highly had a better chance of thriving in old age, with the research finding that 'mental ability . . . at age 11 was a significant predictor of functional independence at age about 77.'[9]

If that indicates the predictive value of cognitive testing, it still leaves the question of what exactly is being assessed when we are given rows of circles, dots and squiggles and asked to pick the pattern. Not to mention the impact of all the social-economic factors at play (remember this was a test of every single child in the country). You would assume that these puzzles have been carefully designed to test how well our brains work at comprehending new information, solving problems, identifying trends and performing verbal or numerical reasoning tasks on the fly (what psychologists define as fluid intelligence, as opposed to crystallized intelligence which relies on using existing knowledge). Yet studies have shown that, under certain conditions, this kind of intelligence is almost entirely correlated with what is known as working memory, our short-term capacity to retain and process information. That is especially true when the tests are done on a timed basis, at which point fluid intelligence and working memory become 'statistically indistinguishable' (whereas without the constraint of time, the correlation drops to around a third).[10] In other words, when working under pressure, the test is not so much measuring how smart we are as it is how good we are at holding information in our heads. Are you clever or do you just have a good short-term memory? With the time constraint in place, an IQ test can't necessarily tell the difference.

The neuroscientist Professor Rogier Kievit, who drew this point to my attention, explains: 'Under conditions of high time constraints, that type of abstract reasoning essentially becomes a working memory task. How quickly can you extract the features, can you maintain them in your mind and essentially, given the high degree of time constraint, make an educated guess?' Whereas, without the time limit: 'You have the time and the opportunity to generate possible patterns and generate hypotheses and say if this is the pattern then I should observe this, or no I don't see that, let's go back to the drawing board.'[11] As this shows, one test becomes two different tasks depending on the conditions under which it is conducted. This is a problem scientists must constantly be conscious of as they design experiments and collect data. It's easier than it looks to be measuring something different in practice from what you set out to record in principle. The variables and contextual factors you hadn't considered can quickly skew the results you were hoping to capture.

It is not just the context of an experiment and how it is designed that may affect the outputs, but the nature of the testing group. At the same time as Scottish schoolchildren were sitting their IQ tests in 1932, the Russian psychologist Alexander Luria was conducting some parallel experiments of his own, putting people in Uzbekistan – then part of the Soviet Union – through their paces. He quickly discovered that, while the psychometric tests he was using had been designed with a particular purpose in mind, his subjects – mostly illiterate, living and working rurally, and with no education – interpreted them quite differently. In a classification exercise, given a series of objects and asked to pick the odd one out, respondents largely declined to engage with the concept of grouping similar items together. Shown pictures of a hammer, saw, log of wood and hatchet (small axe), one subject suggested: 'They all fit here! The saw has to saw the log, the hammer has to hammer it, and the hatchet has to chop it . . . You can't take any of these things away.' And another: 'If one of these things has to go, I'd throw out the hatchet. It doesn't do as good a job as a saw.'[12] With a different set of pictures – knife, saw, wheel, hammer – one sixty-two-year-old, when prompted with the suggestion that three of the objects were tools and one was not, retorted: 'But you can sharpen things with a wheel. If it's a wheel from [an ox cart], why'd they put it here?'[13] Whereas the psychologists who had designed the test saw the task of classification as an abstract reasoning task, for Luria's Uzbek peasants it was one of practical necessity – not how do these objects fit together in theory, but how they do work together in reality? My autism also makes me prone to this in that I often take things literally and categorize differently – for example like when I tried to open a door using a light switch on a third date. Through the lens of their life experience, the relationship of a saw to a hammer and a log only mattered as far as one could be used to work on the other. What would be the purpose of seeing it any other way? Luria confirmed that this was largely a matter of educational level by including a group of younger subjects who had been at school for a year or two and tended to classify the objects into conceptual groups, in the way the experiment had intended.

Luria's work is a vivid (and very funny) example of how the results coming out of an experiment can differ markedly from the intentions that went into it. A group of psychologists may deem that image classification is a good way of measuring intelligence, but what does that test tell you about people who have no use in their lives for abstract reasoning, and have learned to make associations differently? Can someone 'fail' a test whose premise they either reject or fail to recognize? Where does neurodiversity lie in this if we are wired differently? 'The same task can mean different things to different people,' Rogier Kievit suggests. 'If [adults] do three times six, for us that's a memory task, but for a five-year-old it's an arithmetic task, because they have to perform those operations. So even the same thing can mean something different to different populations.'[14] Yet when unexpected results emerge, a researcher may actually learn something from an experiment that has gone 'wrong'. While Luria's subjects did not engage with the classification exercise as expected, their responses were revealing, showing the heterogeneity of people's thought processes and how these depend on so many contextual factors. These kinds of results can be the most valuable of all, forcing the researcher to question their experiment and the assumptions that went into designing it.

As these examples underline, scientists must spend as much time thinking about how they collected a certain set of results – in what conditions, under what constraints, from what subject group – as about the implications of the findings themselves. No data set can ever be taken for granted. The researcher must ask themselves whether it is relevant to the question at hand, how comparable it is to other extant data, what its potential flaws and biases are, and how easy it would be for others to replicate. This work to translate and interpret information, holding it up to the light to ask what it does and doesn't tell you, is the bread and butter of science. Anyone can take a stray data point here or a single study there and draw dramatic-sounding conclusions. Sometimes an experiment will produce almost no viable results. (I actually knew a fellow researcher who managed to construct a whole section of their thesis from a single data point. Hard times.) But the real work of science is to dig past the headlines, to be honest about anomalies (are they pure outliers, or an important indicator of diversity?) and hone a genuine impression of what you have or haven't learned. To see what is actually there – or, if you can't, to find a different method for extracting the information that you were looking for.

So, think carefully about your data and beware the risks of both over-interpreting and over-complicating. It always pays to do as a scientist would, and go beyond the immediate data point of how you are thinking and feeling about something in the moment. A person might be a legitimate long-term relationship prospect, or just someone who gave you a bit of what you needed in a particular moment. (Both are good, but it helps to know the difference!) Maybe you've felt like changing job all year because it's the wrong one for you, or perhaps feeling unhappy at work is linked to some other part of your life? Probe the circumstances, question the cause and put the result into its proper context if you want to understand it.

Those are just some of the challenges for the scientist of translating the information in front of them. But interpreting what we can observe is just one part of the scientific puzzle. When we turn to more experimental parts of research, such as theoretical physics, the problem goes from making sense of what you can see to intuiting what you cannot. The job of interpretation evolves from being an honest broker of evidence to becoming an explorer of intangible concepts: reaching for ideas whose existence is fragile and hard to frame, but which have the power to change how we understand the foundations of our world.

That is the challenge faced by astronomers who, as they have been able to study the universe in ever greater detail, have arrived at a problematic conclusion: what you see is very far from what you get. Even with the most powerful telescopes, vast swathes of the galaxies that surround us remain invisible. The mass that they are made up of cannot be seen – its effect can only be perceived indirectly. This is what we call dark matter, a mysterious entity that sits at the end of several brilliantly twisty branches of science: where quantum theory meets theoretical physics and cosmology. Here things get wonderfully complicated and interpretation becomes not just a big part of the game, but the whole kit and caboodle. Because researchers in this sphere cannot reach for tangible evidence, they must develop their theories and conclusions around a picture with huge pieces missing. They can only fill those gaps by making assumptions about what we can't see based on the behaviour and response of what we can. It's exactly this process of deduction and suggestion that has led theoretical physicists to a remarkable conclusion: almost all of the universe – a whacking great 95 per cent – is and may never be visible to the scientific eye. But we know it's there, and we have an increasingly good idea of what it's up to.

So how did scientists go about discovering something they couldn't see? For our purposes, the story begins with Vera Rubin, a pioneering astronomer who changed how we understand the world while shrugging off sexism at every turn: after she secured a college scholarship, her high-school teacher encouraged her to 'stay away from science'; when she completed her thesis, her supervisor tried to present the findings at a conference under his name; and when trying to use the cutting-edge equipment at the Palomar Observatory in California in the 1960s, she was told that it was not suitable for her since there was no women's toilet.[15] Unperturbed, she went on to produce defining work in her field. Notably, in 1968 she conducted a study of Andromeda, the nearest galaxy to the solar system. She and her research partner were examining the movement of stars in Andromeda and what is known as the 'rotation curve': the velocity (speed and direction) at which stars move relative to their position in the galaxy. Newtonian gravity tells us that the further away from the centre (and source of gravity) an object is, the slower it should be moving – or opposingly, the closer an object is to a source of gravity the more it gets propelled inwards. The reduced gravitational force being exerted on it should slow its acceleration in turn.

Yet Rubin did not find this 'downward' rotation curve of stars moving progressively more slowly. Almost immediately, she saw that the curve was in fact flat: however far off the stars, they continued to move at the same velocity. The distance was increasing, but the apparent gravitational force was not diminishing as a result. Something else must be out there, dictating the apparently unnatural movement of these distant stars. For the stars to be moving so fast, there had to be more gravity, which meant there also had to be more mass.[16]

It was the prelude to a remarkable realization: that a significant amount of mass in the universe – in fact, as we now understand, the overwhelming majority of it – could not be seen or observed in any way. Its presence was the only way to explain what astronomers *could* observe, but this was a matter of deduction and not tangible proof, the result not of direct evidence but what the gaps in the story implied. As Vera Rubin said, accepting this idea required setting aside one of the fundamental assumptions not just of science but of life – that seeing is believing. 'Nobody ever told us that all matter radiated. We just assumed that it did.'[17]

Thirty years after Rubin conducted her study of Andromeda, cosmologists overturned another consensus about the universe. They had long understood that it was continuing to expand, as it is doing right now while you read this. And they had thought there would be an obvious consequence of this: the ever-increasing mass of the ever-expanding universe would create a gravitational drag, *slowing* the rate of that expansion. By taking this idea to the theoretical extreme, it was hypothesized that the collective mass would eventually become so great that the whole thing would collapse in on itself – a drag turning into a slingshot, in a cosmic finale known as the Big Crunch.

By the 1990s, researchers had identified a way of measuring this pace of expansion with some accuracy. By studying supernovae (exploding stars) known as white dwarves, they could track their brightness across different distances, and so infer the speed at which they had travelled. Like Rubin and the rotation curve of stars, the cosmologists who studied this question had a clear expectation in mind, believing that the data on relative brightness would verify the assumption of a universe expanding more slowly. Yet when the numbers came back, they exploded this convention in an instant. When one of the researchers used this data to try and complete an equation that would equate the presumed rate of expansion to the combined mass in the universe, it spat out a negative number for mass. In other words, the equation was the wrong way around. The universe was not decelerating but in fact speeding up. And not just a little bit. A competing team (two had been researching the same issue in parallel) found that the most distant supernovae it studied were fainter (i.e. further away and had travelled faster) than could be explained even if the entire universe was completely devoid of mass![18] Which is the theoretical physics equivalent of learning that Santa doesn't exist. Even in a model with *no gravity at all* to hold them back, these stars should not have been moving so quickly. The only explanation was that some opposite force must have been at play, 'a new component of the universe with negative pressure, [causing] the repulsive variety of gravity to dominate over matter's attractive gravity'.[19] This force was effectively cracking the whip to make the cosmos go faster, when Newtonian gravity should have been pulling on the reins. A new term was coined for it: dark energy.

In the decades since the existence of dark matter and dark energy was first inferred, the search for them has become all-consuming, and continues today with ever more sophisticated tools and technology. Scientists are going to extreme lengths, literally, in pursuit of these invisible mysteries: one of the most sophisticated labs is located in Canada over 2 kilometres underground, to shield its sensors from radiation that would interfere with their readings.[20] A key observatory is located on the South Pole, where optimum conditions for astronomy are created by the absence of sunlight for six months of the year, the relentless dryness and almost imperceptible winds.[21] Yet these elaborate searches are still failing to pin down what theoretical physicists regard as the building blocks of our universe. Cosmologists are convinced their Loch Ness monster is out there, but they still haven't caught a glimpse of it. It's an almost perfect form of scientific torture. The stuff that makes up an estimated 95 per cent of the universe (68 per cent dark energy and 27 per cent dark matter), which could hold so many answers about the evolution of our world and everything in it, is keeping coyly out of view, like a scared kitten hiding behind the sofa. Scientists can perceive the effects of dark matter and energy, they can talk knowledgeably about their role and impact, magnitude and likely location, but they can't actually see – let alone isolate or directly study – the stuff. As Richard Panek, author of a book on the subject, has written, being a student of dark matter means 'coming to terms with a deep irony: it is sight itself that has blinded us to nearly the entire universe'.[22]

However elusive they may be, the search for dark matter and dark energy has a lot to teach us about how scientists interpret information – whether making sense of unexpected data or pursuing a theoretical concept that eludes concrete reality. These areas of physics show the need, when straightforward observation is not possible, to look past what is absent and see what can be evidenced – to let an idea emerge in the negative space of what is left when everything tangible has been taken into account. In these circumstances, our innate human ability to ask questions and make deductions based on what *isn't* there and *can't* be measured becomes more important than ever. Scientists live and die, in other words, on their ability to interpret based on limited information: to piece together a puzzle even though the dog has chewed half the pieces in the box. Dark matter may not be visible, but that does not mean it leaves no trace, the theoretical physicist Sean Carroll told me: 'It affects the stuff around it. It affects the light that is passing through it, it affects the matter that is congealing around it, and that is a lot of observational constraint. There's not a lot of wriggle room. Once you put all that data together, you can more or less figure out how much there is, where it is, what it's doing.'[23]

Knowing that you have to make deductions and draw inferences is one thing. Actually trusting yourself to do it is quite another. It can be doubt inducing when an experiment shows a finding that appears to contradict some fundamental part of received scientific wisdom. During the work that led to the discovery of dark energy in the 1990s, it was an astrophysicist called Adam Riess who performed the fateful calculation that showed people were wrong about the universe's speed of expansion. It was his computer that showed a negative number where a positive one had been expected, implying that things were moving faster and not slower. Not only was this entirely contrary to what established physics told him to believe, but it implied that an idea Albert Einstein had first adopted and then abandoned as his 'greatest blunder' – the cosmological constant, effectively a form of reverse gravity that repulsed matter – might actually exist. Having been disowned by history's most famous physicist, the idea had effectively become unmentionable in the field – a kind of first rule of physics *Fight Club*. The finding that it might actually exist could not simply be written up and unleashed willy-nilly on the world. This wasn't just a case of turning the machine off and on again – the unexpected number had to be carefully scrutinized and verified. '[Even] my modest experience told me that such "discoveries" are usually the result of simple errors. So I spent a couple of weeks double-checking my results but could find no errors.' He then had a colleague – who described his reaction to the result as 'somewhere between amazement and horror' – double-check the work independently. Finally the pair systematically went through and discounted the reasons that it might be wrong: that the most distant supernovas, 'born when the universe was younger, might somehow be different'; that there was undetected space dust interfering with the brightness readings; or that the work was flawed because astronomers are biased towards identifying and studying the brightest objects in a galaxy.[24] In this way, unlikely findings and improbable conclusions harden from wet clay into hard-baked discoveries – by checking and re-checking the numbers, considering alternative theories, and thinking of every possible reason people might put forward to say that you are wrong (which scientists have to do all the time, explaining why we're all such humble beavers). Slowly, sceptically, the impossible starts to become thinkable. The same had been true decades earlier with Rubin's research into galaxy rotation curves: 'We thought that Andromeda was a peculiar galaxy, and that the next one would be more like what we expected,' Rubin later recalled. Only after the pair had studied dozens more did they start to believe what they were seeing.[25]

Just as scientists sometimes have to poke about in negative space to find what they are looking for, they also have to rummage around in the dressing-up box of alternative explanations when they stumble across something improbable. Only by chucking out all the costumes that don't fit do they gradually become comfortable wearing one that initially seems to clash so badly with the established consensus. 'As a sanity check, you say despite the fact it's fitting the data perfectly, maybe I'm wrong anyway,' is how Sean Carroll describes this part of the process, in relation to dark matter. 'Maybe it still is the theoretical side of things that has gone wrong. It's not that there's more matter out there but my theory of gravity was wrong. So you try to do that: you try to invent different theories of gravity, modifications of the known theory that would explain this . . . but the theories just don't work as well, they don't fit the data as well, they're not very elegant, they're not simple.' It is through this kind of iteration, he says, that dark matter has established itself as the new orthodoxy: 'We still haven't pinpointed from an experiment point of view what the dark matter is. But for decades now the "fit" of the dark matter idea to a whole series of observations has just gotten better and better. Whereas alternatives have just gotten worse and worse.'[26]

Looking at how scientists interpret evidence provides a window into the strange combination of optimism and pessimism, confidence and uncertainty that is part of any research effort. A good scientist needs to be convinced enough of the validity of their hypothesis to pursue it hard, but not so convinced that they fail to take account of contrary evidence as they go along, or fall into the 'pathological science' trap of proving what they want to find. And they need to be sceptical enough about the data they find to identify its shortcomings, without being so sceptical that they overlook the possibility of having discovered something remarkable and consensus shattering. Like so many things in science, interpreting the results of an experiment is the search for a sweet spot: neither torturing nor massaging the evidence into saying something it does not really show, nor flattening something that is messy and interesting into fitting with what you had initially expected. It is both an uncomfortable sport and an art in itself.

Nobody knows exactly how to balance these two things: when to stick and when to twist. When Adam Riess shared with his colleagues the data that showed the long-denounced cosmological constant might in fact exist in some form, a lively debate followed (over email) about whether to press ahead and publish. Most were sceptical: 'How confident are we in this result? I find it very perplexing'; 'We all know that it is FAR TOO EARLY to be reaching firm conclusions about the value of the cosmological constant'; 'I am worried. In your heart you know [it] is wrong, though your head tells you that you don't care and you're just reporting the observations.' Riess's own response was a beautiful encapsulation of what science should be about at its core. 'The data require a nonzero cosmological constant! Approach these results not with your heart or head but with your eyes. We are observers after all!'[27]

Just like anyone else making a big decision, scientists can overthink things, and get tied up into the social binding of performance and practice. They worry about getting it wrong in public, the impact on their professional reputation, and whether it will stop an experiment being published or affect the funding needed to continue their research. They can spend so much time thinking about the ramifications of their work that they risk losing sight of the work itself – forgetting, as Riess suggested, that the essential task is to consider information honestly and open-mindedly, however surprising the results and however risky it may feel to put them out into the world. As they move from observation to experimentation and start to collect data, scientists cannot lose the willingness to be surprised by what they see, or their appetite to grapple with the implications of something unexpected.

In science, as in life, interpretation is a blessing – the urge to be critical and sceptical has prevented many an error and poked holes in lots of cosy consensuses. We should never take what we see or are told as read. But nor can we ponder ourselves into paralysis, twisting every potential implication around in our head until we no longer know which way is up (guilty). Once rigorous checks have been made, potential sources of error been accounted for and alternative explanations duly considered, the time comes for action, however uncomfortable. If you keep on interpreting and re-interpreting what you have found, it will eventually lose all meaning and form, like a mouthful of food turned to pulp. You can't window-shop for ever: whether for scientific ideas, a career move or your dream home. At some point you have to stop doing preparatory work and start taking action: to decide that this is where you want to live, now is the time to quit your job, or this is the experiment you are going to pursue. You will rarely have complete information and by definition there will always be possibilities you haven't considered. But big decisions in life only matter once they have actually been made. Sometimes you have to gather up your courage, trust the evidence of your own eyes, and stop chewing.

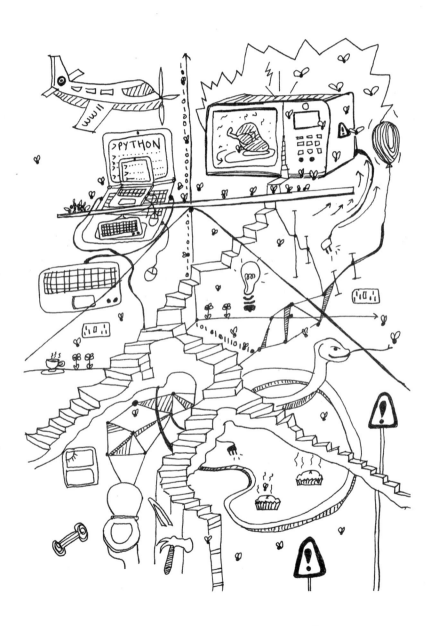

5. Error, Failure and Troubleshooting

How to go wrong and get it right

I never tried my hand at DIY until I really had to. It was after I had the obviously great idea of standing on the toilet to reach a window I didn't usually open. The results of this experiment were clear enough: me on the floor, my ankle bruised, the toilet seat broken and the window untouched. I knew what was supposed to happen next: call someone in to fix it. Because, although I have some fancy letters after my name, none of them qualify me for handiwork. I'm more suited to squeezing a pipette than wielding a hammer, to tinkering with an algorithm than adjusting a wingnut. Or that's what I was used to telling myself. The difference this time was that we had just been through months of pandemic-induced lockdown. This enforced detachment from society did all sorts of things to us, one of which was to weaken our attachment to the habitual way of doing things. When you literally can't call anyone in to help, you learn either to fix things yourselves or leave them broken. And by now I was the veteran of a solo-constructed IKEA bookshelf, put together with the help of only a dumb-bell and pair of tweezers (don't ask).

So as I sat on my bathroom floor, nursing my injured ankle and my pride, I realized that, even though I could have called for help, I didn't want to. Instead, I hobbled my way down the high street to a DIY shop (Robert Dyas, who quickly became the husband I never knew I'd needed), found what I was looking for (a seat that, rather wonderfully, was named Constance) and – after lots of self-affirmation and too many YouTube ads – managed to detach the broken seat and fix a new one. My prize was more than a repaired throne on which to sit. It was also a boost to my confidence, and a reminder of the importance of veering off course to try something we normally wouldn't, risking embarrassing failure in the process. That isn't just good life advice, pushing ourselves to try new things and embrace the unfamiliar. It's also intrinsic to good science: so many discoveries have been the result of dancing with defeat – attempting something that others said wouldn't work, persevering with an experiment that appeared to be fruitless, or finding a creative approach that departed from the textbook. Errors, failed experiments and disproven hypotheses are fuel in the engine of any scientific inquiry – by showing you what doesn't work and (hopefully) why, they move you one step closer to devising something that does.

A paradox of scientific research is that it's based on proven concepts and tested theories while sometimes requiring its participants to break or cast aside those rules. Scientists may have to ask themselves what would happen if a principle that everyone thinks to be true actually isn't. They sometimes have to keep going despite all evidence suggesting that they are cycling down another cul-de-sac. And they can be compelled to completely reverse course and change the parameters or objectives of an experiment if a more interesting opportunity presents itself – a possibility that simply has to be tracked down. There is no instruction manual for how to break the rules as a scientist. The 'right' course of action can entail persevering for years or decades in the wilderness, or doing a 180-degree turn within the course of a week. Such decisions, of course, are clouded in uncertainty. You never know for certain whether to stick or twist – if a pivot is going to point you in a better direction, or just leave you on the floor with a twisted ankle. You don't always know whether to do what instinct is pushing you to, or to try something even when it goes against what every fibre of your brain feels comfortable with. The only certainty is that you will sometimes get it wrong, and embracing that reality is a more productive attitude than trying to avoid it.

As the theoretical physicist Chiara Marletto suggested to me, scientists must embrace the philosophical concept of fallibilism – that some of their beliefs will be wrong and there is no such thing as absolute certainty, only a constant process of inquiry to acquire knowledge. 'It's where you are comfortable with the idea that errors will be made, but problems can also be solved. So whenever you make a mistake, it shouldn't be considered a bad thing, but an opportunity to fix something.'[1] (Exactly what I told myself after breaking the loo seat.) It's also something I've found as a computational biologist, teaching myself the programming language Python from scratch. When you are still learning the ropes, every error and bug can feel like a massive failure, proof that you are not actually capable of doing this. You risk going from being invested in the process to being a hostage of its outcome. But with time, patience and practice, I am now confident enough to know that a bug isn't so much a giant red squiggle across my work as a clue, enlightening me to something I haven't yet seen. The failure isn't the end, just another turn in the road.

But this is where things get tricky, because while humility and the acceptance of failure can be important assets for a scientist, there are times when the opposite attitude is required. Important things have been achieved by researchers who weren't either flexible or humble, who stamped their feet, insisted that their way was the right way and they were going to prove it no matter what anyone else thought.

Some were not even recognized within their own lifetime. Consider the case of hand washing and the role of hygiene in preventing infections from spreading. Today we would call that common sense, but the idea was not proposed until the late 1840s, when the Hungarian doctor Ignaz Semmelweis, while working in the maternity wing of a Vienna hospital, set out to understand why mothers were dying of fever at a much greater rate on some wards than others. One of his experiments was to make hand washing with disinfectant mandatory, which quickly had the effect of reducing mortality, since the doctors treating patients in the high-risk wards had also been handling corpses to conduct autopsies. But far from being thanked for this work, Semmelweis was effectively driven out of his job and continued to be opposed by the medical profession. (One colleague in Vienna said he had got the wrong end of the stick, and deaths had gone down because the hospital had improved its ventilation.) Not helped by his pugnacious attitude towards other doctors (threatening to declare one a murderer, and describing another as a 'Medical Nero'), his work went unrecognized during his lifetime and he died aged forty-seven in an insane asylum. Only posthumously did this very stubborn scientist earn his legacy as the 'father of infection control'.[2]

We owe many important discoveries to scientists who, like Semmelweis, stayed loyal to their ideas even when they were ridiculed, and when there was a personal or professional price to be paid. As Professor Frances Balkwill of the CanBuild cancer research programme told me: 'It's a real problem for a scientist who can't see the blind alley. But of course some of the great discoveries have come from people who said, "I do believe this is going to work." . . . Some of the greatest inspirational stories [are] people saying, "That's *not* a blind alley."'[3] Just as we wouldn't have some of our most famous books without authors willing to carry on in the face of rejection – *Gone With the Wind* was rejected by almost forty publishers, *Moby-Dick* was deemed too long and *Twilight* was turned down by fourteen different agents – science would be immeasurably poorer without the bloody-minded researchers who insist that they are on to something, and pursue that idea until they have the evidence to persuade others.

So failure is a tricky frenemy in the world of science, as in life. Everyone is stalked by it, but there is no universally correct response to it – apart from rolling with its punches and accepting it as part of experimental design. There's a time to accept failure and a time to fight it. A time to pivot and a time simply to walk away. Science can't give you the perfect answer about how to respond to these dilemmas, but it does shed light on our relationship with failure as humans, and the different ways to handle it. And it underlines one important lesson: failure, whether you have anticipated it or not, can be one of the best teachers, pointing the way to a better approach or highlighting a previously unseen flaw. Often improvement only comes once some kind of error has emerged to show us what we've been needing to fix. Sometimes a wrong turn is exactly what you need to (eventually) get it right.

The Covid-19 vaccine programme was astonishing in both its speed and its scale: within two years of the first dose being administered, to ninety-year-old Margaret Keenan, more than 5 billion people around the world had received one, over two-thirds of the global population. At the outset of the pandemic, many would have guessed that vaccines would still be in the early stages of development at this point: no equivalent immunization had ever been developed in less than four years. Yet they were in fact close to completion before most people had even heard the word 'coronavirus'. By January 2020, when a pandemic was still two months away from being declared, scientists had already mapped the genetic structure of the SARS-CoV-2 virus, and within weeks a prototype vaccine followed.

The ability to move so quickly was not just a reflection of the power of modern data science, with processes that would once have taken years now being completed within hours. It was also the product of years of research in an apparently unpromising area of molecular biology, which would become familiar to millions of people, but at first was largely dismissed by the select few who had studied it. The sprint to develop and test the Covid-19 vaccines in a matter of months was only possible because of the decades of research into the initially obscure field of mRNA – a story that underlines how dramatic scientific success can be built on a foundation of relentless failure, and the debt we owe to scientists who doggedly stick to their guns when everyone is telling them to give up.

The real story of the Covid vaccines began not in the wet markets of Wuhan, or the Coventry hospital where the first dose was administered, but by a photocopier at the University of Pennsylvania more than twenty years earlier. Two scientists were queuing to use the machine; unknown to each other, both were holding jigsaw pieces that would prove to be an awesome combination. Dr Katalin Karikó had spent years studying messenger RNA – the biological molecule that transmits information from DNA, instructing cells in the body to produce the proteins that carry out their work (such as repairing tissue, generating immune responses and producing hormones). DNA may hold the code that dictates how every living organism develops and lives, but it is RNA that reads the recipe book and calls out orders in the kitchen. First identified in the early 1960s, it had long been the subject of interest for an obvious reason – if this was the molecule that acted as the cell's control centre, then the ability to replicate its function offered massive clinical potential, such as getting cells to produce targeted antibodies that the body didn't know how to make itself, or to grow different kinds of tissue. In theory, it was the gateway to an untold wealth of medical possibilities. Yet its potential with regard to vaccines had been explored only sporadically in the intervening decades. That was where the second scientist waiting by the copier, Dr Drew Weissman, came in: he was an immunologist who was then researching an HIV vaccine. Dr Karikó, whose focus had previously been on therapeutic uses of RNA (such as training the brain to produce more of the chemical that prevents blood clots), agreed to work with him on developing an mRNA-based approach.[4]

It was the pair's subsequent partnership that helped to turn mRNA from a fascinating possibility into a tool with tangible applications. In theory, the idea of developing mRNA for immunization was coherent: what better way to help the body combat a particular virus than to mimic its own system for generating antibodies? Yet in practice it proved endlessly difficult – a reminder of how much failure, trial and error, and adjustment is involved in any major scientific undertaking. Researchers had been messing around with RNA for decades by this point, with breakthroughs including the successful creation of synthetic mRNA in 1984, and the discovery three years later that it could both promote and inhibit protein production in cells. By the early 1990s, it had been successfully trialled with mice, generating an antiviral immune response. Yet at the time of the photocopier summit in 1997, mRNA was seen less as a potentially transformative area of research, and more as a probable white elephant: expensive, tricky to work with, and not worth the candle. Several ambitious projects had been abandoned, including an attempt to develop an mRNA vaccine for flu. It was deemed expensive, unstable and generally unsuitable for large-scale investment. Few seriously believed that a medical revolution was afoot: 'If you had asked me . . . if you could inject mRNA into someone for a vaccine, I would have laughed in your face,' one pharmaceutical executive who supplied equipment to RNA labs reflected.[5] Such scepticism was widely shared. After molecular biologists at Harvard first created synthetic mRNA, the university declined to patent the process, which continues to be used today. Some onlookers were downright contemptuous: when Ingmar Hoerr, co-founder of the RNA company CureVac, presented research on mice trials in 2000, a Nobel Laureate stood up in the front row to shout him down.[6]

Even among true believers, there was plenty of good reason for doubt. Katalin Karikó's work had consistently been running into one formidable stumbling block. Although she was able to manufacture mRNA molecules in the lab and get the desired response from cells *in vitro*, it was a different story once those same molecules were introduced into a live body. Time after time, the synthetic mRNA would provoke a concerning immune response. 'Nobody knew why,' Dr Weissman remembered. 'All we knew was that the mice got sick. Their fur got ruffled, they hunched up, they stopped eating, they stopped running.'[7] A body that quite happily produced its own mRNA molecules was striking back against the synthetic equivalents that scientists had produced, with its immune system going into overdrive and serious inflammation resulting.

The solution to that problem would sow the seeds for mRNA vaccines to enter the mainstream. Through a control in one of their experiments, Karikó and Weissman observed that a different form of RNA (transfer RNA) did not lead to inflammation. By isolating a molecule within the tRNA and adding it to their synthetic mRNA, they were able to avoid the immune reaction. This was modified mRNA, the basis for the successful vaccines that still lay over a decade in the future. It was a remarkable invention: the body did not just accept it, but responded far better, with cells producing much more of the desired protein in response.

It was a huge breakthrough, but not – at least initially – recognized as such. As Dr Weissman recalled, most of the grant applications the duo made based on modified mRNA were rejected. 'People were not interested in mRNA. The people who reviewed the grants said mRNA will not be a good therapeutic, so don't bother.' They also struggled to get published in scientific journals or to gain interest from industry: 'We talked to pharmaceutical companies and venture capitalists. No one cared . . . We were screaming a lot, but no one would listen.'[8] Despite promising test results – including an experiment that found mRNA could help monkeys to significantly boost their production of red blood cells – most doors were slamming shut in their faces. Even the eventual publication of a paper documenting their success with modified mRNA failed to shift the dial. 'I told Kati [Karikó] our phones are going to ring off the hook. But nothing happened. We didn't get a single call.'[9] This all helps to show that science isn't always a fair fight: having the evidence on your side isn't necessarily sufficient to get the attention and funding needed to progress an idea further. Sometimes the error isn't in the research, so much as in the willingness of the wider world to recognize or embrace it. Overcoming scientific error is one half of the challenge; the other is getting a sometimes reluctant system to listen. (Climate scientists, this one's for you.)

In the case of Karikó and Weissman, it was not until several years later that their work finally gained the attention of two minnows in the pharma industry which the Covid-19 pandemic would turn into unlikely household names: Moderna and BioNTech. Both companies licensed patents from the two scientists and supported their work – investments that would prove crucial as the novel coronavirus began wreaking havoc at the beginning of 2020. After more than half a century of research, this was mRNA's moment in the sun, and it did not disappoint. The design of a coronavirus means that it uses a sharp point on its surface – the spike protein – to inject itself into cells and spread across the body. The Moderna and

Pfizer-BioNTech vaccines work by injecting mRNA molecules that instruct the cell to produce a mock spike protein, effectively training the immune system to recognize and repel the real thing. It was the long-established promise of mRNA – to put medical science at the controls of the body's protein factories – brought to dramatic fruition. When the data returned from comprehensive human trials, the efficacy of these vaccines was shown to be in excess of 95 per cent.[10]

The successful development of Covid vaccines, at unprecedented speed, exemplified many important truths about the scientific process. It showed how it can be painstakingly slow and intimidatingly long term, with the first glimmer of a possibility sometimes taking decades to mature into a real-world application. Synthetic RNA was well into middle age by the time scientists were able to deploy it in a clinical setting. It also underlined the many hands that are involved in what appears to be a singular breakthrough. Karikó and Weissman's contribution was significant, but they were just two of numerous researchers who worked on mRNA before and alongside them, reflecting how science so often proceeds as a patchwork of projects and initiatives, each adding puzzle pieces to the pile, making the full picture that little bit clearer in a system ready for it.

Perhaps most importantly, the mRNA vaccine demonstrated that scientific success can – and perhaps must – be the child of failure. For most of mRNA's history as a recognized scientific concept, any reasonable onlooker would have concluded that efforts to unlock its clinical potential were failing: as we have seen, many scientists and industry professionals did exactly that. The test results were unpromising, the material was notoriously difficult to work with, and funding was painfully thin on the ground. The principle may have been great, but the implementation of it was looking stuck somewhere between improbable and impossible. It was not just the science that struggled, either.

Katalin Karikó saw her career suffer as she pursued her vision for mRNA. Two years before meeting Weissman, she had accepted a demotion and a pay cut to continue her research in Pennsylvania – a woman true to her cause. It would have been easy to have followed another path, to have accepted what so many people were saying and gone where the grant money wanted to go. Instead she stuck to her guns – not just accepting failure but actively courting it in her pursuit of mRNA treatments. As one of her former mentors, Dr David Langer, has said: 'There's a tendency when scientists are looking at data to try and validate their own idea . . . The best scientists try to prove themselves wrong. Kate's genius was a willingness to accept failure and keep trying, and her ability to answer questions people were not smart enough to ask.'[11]

Without this attitude, refusing to be intimidated by failure and instead working hard to learn the lessons of each unsuccessful experiment, it is conceivable that the mRNA vaccine would not have been a weapon in science's arsenal when SARS-CoV-2 reared its head. Instead, an idea that was repeatedly denounced by informed experts became a vital part of the fightback against a global pandemic. There could be no better example of how scientific progress is built on a foundation of ideas that didn't work and experiments that appeared to fail. Because the underlying premise of mRNA was robust, the researchers who truly believed in it were able to keep going despite the widespread criticism, the lack of financial support, and the sometimes concerning test results. The final vindication came in 2023, when Karikó and Weissman were jointly awarded the Nobel Prize in Physiology or Medicine for their work that had enabled the development of mRNA vaccines.[12] A stellar example of how taking failure on the chin, sticking to your guns and rolling with the punches pays off.

It's an example we could all do well to learn from. When

something really matters to us, and we truly believe in it, external criticism or initial stumbles shouldn't in themselves be enough to dissuade us from pressing on. There is, as the scientist and philosopher Professor Massimo Pigliucci suggested to me, always more than one route to take in a situation where the outcome is uncertain, and you are unsure about the next step. 'Some people seem to think there is one reasonable thing to do, or only one reasonable course of action, but that is not true. There may be multiple reasonable courses of action for any specific problem . . . There are also a lot of ways that are not reasonable. The trick is to figure out, are you on one of the reasonable paths here, or are you actually going in an unreasonable direction?'[13] It's good advice for those of us who fill journals and endure sleepless nights trying to think about the 'right' thing to do, when science is telling us that it probably doesn't exist.

When we are trying to do anything difficult in our lives, it's important to remember that there is rarely a binary between a good and bad decision, just endless gradations in between and multiple paths branching in different directions. The one thing that unites all this is that failure is inevitable at some point, and you must be willing to face it. Building a business, making a career change, pursuing a passion project or moving abroad are not straightforward things; we shouldn't expect to accomplish them without making errors, feeling like a failure, and getting negative feedback along the way. The lesson from science is not to give up so easily. Instead try to learn from your failures, listen to criticism and adjust accordingly, giving yourself a slightly better chance with each new experiment. Be patient, accepting that what turns out as success may often look an awful lot like failure along the way. Only stop altogether either when there is nowhere left to turn, you have no faith left in the idea, or a better one presents itself. Great ideas and important achievements don't come to fruition easily, or without considerable doubt and dispute along the way. We should all be grateful for those who are dogged and determined enough to see them through to the point of meaningful discovery.

The ability to stick when others want you to twist can be an asset to the scientist, in the right circumstances. And if you are anything like me, hearing someone dismiss your idea before you have had the chance to properly explain or explore it is all the more motivation to make it happen. If I had taken seriously everything my peers and teachers ever said to me, I would never have had the energy and faith to do my PhD, pursue my career as a scientist, and write my books. Let alone assemble a bookcase using 'improper' tools or tackle a broken toilet seat.

But of course stubborn self-belief is not the only attitude that is needed, nor will it always be the right one. Just as important can be the willingness to pivot – to adjust ideas, repurpose experiments and generally show a willingness to abandon prior assumptions when the evidence nudges you in a different direction. Scientific discovery is often not the product of singular thinking, but of threads that were originally spun for one purpose being picked up and unspooled in a new direction, often after a gap of years or even decades. This requires troubleshooting what didn't work, isolating the bits that did, and being creative with what's left.

Tools, technologies and theories are frequently adapted, turned on their head and utilized for a different reason than the one they were created for. That is one of the joys of science: an unruly stampede of research that is ultimately shaped by what has come before – whether to repudiate an existing theory, improve a current solution, or repurpose something altogether. It is about timing, about the question you are asking, and about people's ability to think creatively.

It's for these reasons that you have a microwave sitting on your kitchen counter – an invention that fell sideways out of scientific research that had never even considered cooking, and deserves to be counted as one of the great accidental discoveries (100 million of them are now manufactured every year). The inventor of the microwave – and so the father of the ready meal – was the American electrical engineer Percy Spencer. In 1946, he was experimenting with a distinctly unappetizing device called the cavity magnetron (hot name) when he noticed that the nut cluster bar he was carrying in his pocket had melted. (According to his grandson, he had it with him to feed to the squirrels outside during his lunch break.) Further experiments with popcorn, an egg (which duly exploded in someone's face) and gingerbread mix followed, and the result was the beginnings of the device that would become a household favourite, although you would not have recognized the massive early models, or the first name that was chosen – Radarange.[14]

While this sounds like one of those wacky stories of accidental inventions, the technology it emerged from was deadly serious. The magnetron had principally been developed during the Second World War, as an improvement to radar systems that allowed small detectors to be fitted to aircraft. In very simple terms, magnetrons function by oscillating electrons around a magnetic field, allowing the device to emit the short-length radio waves (microwaves) that are the basis of radar systems – in which the bouncing of waves off solid objects, in this case aircraft and submarines, can be measured to estimate their position. You might think a pivot from spotting submarines to cooking lunch is big enough, but the magnetron had a history even before its wartime deployment. It was in fact first conceived in 1917 by the engineer Albert Hull, of General Electric, who was trying to circumvent a competitor's patent for a device to control electrical current in a valve.[15] Although the idea came to nothing (GE decided to buy out the patent instead), the idea for the magnetron had been born, and would be more fully developed in the decades that followed, culminating in the invention of the much more powerful cavity magnetron in 1940 – the direct forerunner to the element at the back of your microwave today, and saviour of all emergency meals.

Our love for a quick lunch can therefore be linked back to one of the great attributes in science: the ability to pivot, and move a theory or technology in a different direction from the one originally intended. That might be taking something that didn't work, and applying the fundamental idea or process elsewhere. Or it could be the chance discovery of a secondary attribute that opens up a whole new set of possibilities, like Percy Spencer's nutty realization that a device designed to emit microwaves also emanated heat. These are examples of scientific discoveries – and pivots – that are purely opportunistic. Researchers notice something they haven't been expecting, they poke around a bit and they discover something interesting – a loose thread that just demands to be pulled and made into a technology that benefits many.

Not all discoveries are quite so serendipitous. The pivot can also come when researchers engage in one of the essential processes of most experiments: troubleshooting the thing to work out what is going wrong, a step that is almost always necessary. (And let's face it, painful!) Sometimes this is simply a case of establishing that a parameter here or an input value there is off-kilter, and adjusting accordingly. (Which can be easier said than done: at times even dealing with a misplaced comma can feel like an unreasonable distraction from the bigger problem you are trying to solve.) When I am coding I can take errors personally. There are days when you feel defeated before you have even started.

But the troubleshooting process can be much more revealing than finding and fixing errors. Sometimes it takes researchers into even more promising territory than they had at first envisaged. So it proved in 2015 at the University of Chicago, where a team of researchers were working with strontium titanate, a so-called topological insulator (which acts as a conductor at the edges and an insulator in the middle). These materials, first identified in 2005, hold huge potential in the field of computing – where their efficiency as electrical conductors (eliminating resistance because electrons can only travel in one direction) means they could significantly reduce the energy consumption of computers compared to traditional circuits based around silicon transistors. They are also regarded as the basis for the potential breakthrough technology of quantum computing, with its (theoretically) vastly more powerful and flexible modes of calculating data.

The only drawback with topological insulators is their fragility and instability as materials. These are delicate bunnies, which can easily lose their important properties when handled incorrectly (including when just exposed to the open air). Which makes them tricky to work with for potential computing purposes, because the exact same processes we use to create semiconductor devices from silicon (the chips and transistors which power your devices) effectively render the topological insulator useless.

A potential solution to this problem emerged as the product of an unlikely bit of troubleshooting by researchers from Chicago and Penn State Universities. They were conducting experiments on strontium titanate in a specially engineered lab, but kept experiencing 'a slow drift in our measurements' which had been producing noisy data over the course of several months. This meant they had to troubleshoot their experiment, and found an unusual reason for it: the skew in results was being caused by the harsh fluorescent lights in their laboratory. So they switched them off, and had – please forgive me – a light-bulb moment. By performing tests with and without the lighting, they found that the lights were having an unforeseen effect on the material, and a remarkable one. By inadvertently focusing UV waves onto the strontium titrate, the lights were causing it to become polarized. They were managing to forge the electrical circuits that are so hard to produce on this very particular material. Even better, some further tinkering revealed that this process could be reversed by focusing a beam of red light on the surface, making the material reusable. 'It's like having a sort of quantum Etch a Sketch in our lab,' the project's co-lead Professor David Awschalom commented at the time. 'Instead of spending weeks in the cleanroom and potentially contaminating our materials, now we can sketch and measure devices for our experiments in real time. When we're done, we just erase it and make something else.'

Not only was this an important discovery in the field of quantum mechanics, but it also perfectly encapsulated the relationship between failure, error and progress in any scientific inquiry. It was only because the initial experiment was going wrong that the researchers started to troubleshoot – looking for every external factor that could be affecting the results, right down to the room they were working in. And it was only through this troubleshooting process that they glimpsed a new possibility. As Professor Awschalom reflected: 'This observation came as a complete surprise . . . It's one of those rare moments in experimental science where a seemingly random event – turning on the room lights – generated unexpected effects with potentially important impacts in science and technology.'[16] It reminds me of a running joke in my research team: if the code works and runs, no one breathe in case it jinxes it and breaks the spell.

That moment would not have arrived without the happy accident of errors in the original experiment that demanded a closer look. As this shows, troubleshooting doesn't just solve the problem you can identify. It may also create entirely new lines of inquiry, and turn a routine exercise into something scientifically remarkable. For those unexpected breakthroughs, we can thank experimental failure and the way scientists work and persevere to overcome it.

This, perhaps, is one of the more relatable parts of science, even when we are talking about subjects like quantum mechanics that can get very complicated, very quickly. Because, despite the theories, the robust experimental design and the rigorous peer-review process that surrounds it, there is undoubtedly a DIY element to good science: messing around with things to find out how they work, tweaking them to see if that makes a difference, and rummaging about in your toolbox to look for something that might fit. Not quite making it up as you go along, but certainly being willing to go out on a limb, even if that means falling off it.

This is a mentality that good experimental researchers embrace, and you should too – even within the pressures to perform and 'get it right'. It's such a classic human failing to worry about how something will end up before we have even given it a go: to fear some kind of embarrassing failure and use that as the reason to avoid trying. The scientific mindset can be liberating here. It tells us that failure is not the ending, and that failing shouldn't necessarily be seen in a negative light. In fact, it might only be through failure that we learn something important that can change how we think or behave. Maybe you'll find Boxercise wasn't for you, but secretly you liked going to the gym. Or perhaps you won't get the job, but the feedback will help you realize that you're closer to it than you had thought. It's amazing how different the concept of failure feels when you flip it around and decide to credit yourself rather than beat yourself up.

Science teaches us that there is no shame in trying things that might not work – only in failing to analyse those failures, identify the errors that caused them and then equip yourself with that knowledge to try again. That, when you strip away all the technicalities, is really the essence of good science: a willingness to look again when bitten. And if switching the lights on and off to work out what's happening is good enough for highly trained quantum engineers, then it's definitely good enough for the rest of us.

6.   Teamwork

How to collaborate effectively

When news that nuclear fission had been achieved first reached the United States in January 1939, the physicist Luiz Alvarez was sitting in a barber's chair, newspaper in hand. What he read so stunned the future Nobel Laureate that he abandoned the haircut halfway through to run back to his office, the Radiation Laboratory at Berkeley. There he met his boss, the brilliant theoretician J. Robert Oppenheimer, who in the years that followed would lead the programme that turned this scientific discovery into the most powerful weapon known to humanity, the nuclear bomb. All that from an inadvertent asymmetrical haircut (which was surprisingly ahead of its time).

Yet on first encounter with the principle, Oppenheimer was dismissive. 'That's impossible,' he told Alvarez, walking over to a blackboard to chalk up his reasoning.[1] Oppenheimer may be forgiven for this misplaced scepticism. For not even the scientist who had actually conducted the world-changing experiment, German radiochemist Otto Hahn, initially realized what he had done or the full significance of it. Only with the insight and intervention of his long-time research partner, physicist Lise Meitner, did this remarkable penny drop. And so nuclear fission – the splitting of the uranium atom – was not just one of the most consequential scientific discoveries of the twentieth century, which continues to shape our world today. It also offers a prime example of how no scientist ever really works alone, illustrating the fundamental importance of teamwork in the research process, and how collaboration runs through all scientific work, in numerous ways and at multiple levels.

Like most important discoveries it came not out of the blue but as part of wider progress in a particular area of research, in this case the burgeoning field of nuclear physics. Since the final years of the nineteenth century, beginning with the discovery of subatomic particles in 1897, physicists had made great strides towards decoding the structure and behaviour of the atom. Notable breakthroughs had included Ernest Rutherford's deduction in 1911 that atomic mass was concentrated centrally in what we now call the nucleus (sadly displacing a much tastier theory of atomic structure, known as the plum pudding model). In addition, James Chadwick's 1932 discovery showed that, along with protons and electrons, there existed particles that carried no electrical charge: neutrons. After decades of developing their understanding of the atom and its sub-structures, physicists had found a tool that – by dint of not being repelled by a matching electrical charge – would allow them to probe further into the heart of the matter, the nucleus itself. It sparked the chain of experiments that culminated six years later in Hahn and Meitner's discovery, ushering the world into the huge promise and great instability of the nuclear age.

The evolution of nuclear physics underlines that science in practice is the exact opposite of the stereotypical mad professor with curious hair, toiling alone in their lab, cut off from the outside world. As much as experiments can be isolating at times (and a tad hairy), scientists are rarely actually working alone. In reality, scientific research moves forward as a complex web of interrelationships: scientists work directly together, they study under and alongside each other, listen and contribute to each other's work at conferences, and eagerly consume newly published work in search of fresh insights to propel their own (as well as the solid snack selection that is an equally good reason to attend any conference). There is no glorious isolation, only the constant process of experimentation, results and ideas being shared, and a continuous feedback loop in which debates are held about what it actually means (and who should get the credit – it's certainly not all warm and fuzzy). This is the nature of scientific progress in any field: the slightly chaotic process of researchers borrowing, adapting and improving upon the work of others, with the theories or discoveries that result akin to the marble that has rolled through an exquisitely complex apparatus to reach its resting place.

It's something that is beautifully symbolized outside the Francis Crick Institute, where I used to work. I would often have my lunch sitting opposite a sculpture, *Paradigm*, formed of tetrahedral blocks stacked on top of each other, the smallest on the ground and each layer getting progressively larger. To me this captures the collaborative effort that is science: a constantly growing assembly of new theories built on top of the old. Even once an idea has been superseded and is no longer used, it does not disappear, but remains as an artefact showing how we got from there to here – and a reminder that the theories we use today are just another layer of bricks that will inevitably be built over in the future by scientists with better knowledge and more sophisticated tools.

The collaboration symbolized by the sculpture occurs in many different ways. It operates at what we might call the network level – where the scientists involved in a particular area are connected by their shared educational background, overlapping research interests, and the conferences and publications that form the backbone of any scientific career. And it happens at an interpersonal level, in the partnerships and small groups that drive forward the work of a particular lab or research institute. So it was with Otto Hahn and Lise Meitner, the chemist and the physicist whose remarkable friendship and working relationship showed the great power of collaboration in the scientific process, as well as revealing its limitations, with Meitner largely excluded from acknowledgement about the world-changing discovery they had made together, because of her identity as a woman of Jewish origin. Hahn would be awarded the Nobel Prize in 1944 for the discovery of fission, but Meitner never was (even though she was nominated more than forty times across her career).[2]

It was a working relationship that spanned decades, beginning when the pair met at a physics seminar in 1907, two years after Meitner had become only the second woman to earn a physics PhD from the University of Vienna. In the burgeoning field of radioactivity and atomic physics, they worked intensively, co-authoring nine papers in 1908–9. That was despite the gender bar that existed at the time, which saw Meitner frequently treated as a second-class citizen despite her acknowledged excellence as a physicist. At Friedrich Wilhelm University, where both worked, she was not allowed to work alongside Hahn in the chemistry institute. Its head did not permit women to set foot in the laboratory, partly out of a fear that they might set their own hair on fire (another reason I like to let mine dry naturally).[3] Meitner was instead installed in a basement office that had originally been designated for carpentry, not allowed to enter the rest of the building, and only granted access after the prohibition on women attending universities was lifted in 1909. It makes Meitner's career all the more extraordinary that she did so much pioneering work despite being marginalized and having so many barriers put in the way of her ability to collaborate with others.

Unfortunately, more than a century later, many women in science will still be able to relate to her experiences. These days you might be allowed in the room, but it doesn't mean your ideas will be taken seriously, or that you will be listened to in the way a man would be. During one presentation I was asked by a male colleague if the work I had described was actually my own. Although this has been the exception in my career, it reminded me that you are only ever one comment away from being made to feel like an imposter who questions their own place in the lab, no matter how good you are, how much you have read or what you have published.

The fact that it was so much worse for Meitner, and because of her perseverance in the face of overt sexism, are among the many reasons she is my scientific girl-crush. Despite being frequently sidelined, Meitner swiftly progressed her career, at a time when great advances were being made in theoretical physics, and she had come into the orbit of an exceptional generation of scientists. Her social and professional circle in Berlin included five future winners of the Nobel Prize in Physics, one of whom was Albert Einstein. In 1908 she first met Ernest Rutherford, visiting the city as a newly minted Nobel Laureate, and who promptly declared: 'Oh, I thought you were a man!' In its formative period, the world of nuclear physics was a tight-knit community in which the best minds mixed frequently and constantly scrutinized each other's work, looking for ways to harness it in their own research: a timeless example of how scientific discovery proceeds as both a collective and competitive endeavour. Among her circle of pioneering physicists, Meitner mostly remembered the ethos as a collaborative one : 'Each was ready to help the other, each welcomed the other's success.'[4] Despite the harsh conditions of being a woman in science at this time, Meitner's character prevailed as one defined by curiosity, discovery and optimistic resourcefulness – the very essence of what it means to be a scientist. She proved that, even though science may never lose its tendency to marginalize those who think and work in unusual ways, there is ultimately an egalitarianism about it that cannot be suppressed. What you see in the world, and the ideas it sparks in you, matters so much more than what people see in you.

For all the important influences in her Berlin circle, Meitner's enduring scientific partnership was not with another physicist but with Hahn, a chemist. Among other achievements, their early work together was notable for the discovery of element 91, protactinium, in 1918. The partnership subsequently went into abeyance, but would be revived to dramatic effect in the 1930s, following the discovery of the neutron in 1932 and subsequent experiments which showed that bombarding the uranium atom with neutrons could produce new radioactive products: the first lab-based nuclear reactions. For all her individual brilliance, Meitner also embraced collaboration. To produce successful research she knew that she needed her partner, whose very different skills were the necessary complement to her own. 'It was clear to me that one could not get ahead in this field with physics alone. The help of an outstanding chemist like Otto was needed to get results.'[5] (Talk about having it all – Meitner was both talented *and* humble. OK, enough fangirling.) With Hahn and his fellow chemist Fritz Strassmann, she established the Uranium Project in 1934 to explore the possibilities of using the neutron-bombing technique on the world's first-known radioactive element. It was an example of the importance of interdisciplinary research: the painstaking experimental chemistry required to isolate and identify the products of uranium that had been degraded by neutron irradiation, and the theoretical physics that could process and interpret those results, showing how they might advance understanding of atomic structure and behaviour.

This pooling of expertise was most needed when the Project produced its most remarkable result, in 1938. By this time, Meitner had been forced to leave Berlin, the city she had intended to move to for just a few academic terms, but ended up making her home for over thirty years. An Austrian citizen who had been born to Jewish parents, Meitner's already precarious position in Nazi society had become untenable following the annexation of Austria in March 1938. As she faced the almost certain prospect of being dismissed from her role at the Kaiser Wilhelm Institute and forbidden from leaving the country, friends helped to smuggle her to safety in Stockholm, from where she continued her work with Hahn via correspondence.[6] With post being delivered overnight in both directions, this long-distance working relationship allowed them to maintain their partnership, even as both the distance and the political climate conspired to obscure her part in it. It would yield remarkable results, demonstrating the role of collaboration in cracking difficult problems – where scientists with different professional backgrounds and contrasting skills are bound together by a common vision and endeavour.

After four years of work, the real breakthrough came at the end of 1938, after Meitner had been forced to flee Berlin. Hahn and Strassmann had been attempting to replicate an experiment done in Paris, which had found an unknown substance resulting from uranium being bombarded with neutrons. The two chemists observed that the product of the reaction appeared to be behaving like barium, but dismissed this as a possibility: Ba was a markedly different element from uranium, with a much smaller mass, and the consensus at the time was that nuclear reactions could only result in marginal changes to the original element (essentially that you could make chips off the old block, but not split it in two). Instead they believed the substance must be radium (Ra), which was known to be a natural product of uranium decay, and had a similar relative atomic mass. It was Meitner who questioned this conclusion and urged the chemists to look more closely. She had been receiving updates on the work through regular letters, and also spent several hours in a covert meeting with Hahn when he visited Stockholm in November. As Strassmann recalled: 'Her opinion and judgment carried so much weight with us that we immediately began the necessary control experiments.'[7] They duly set out to separate this elusive substance from the barium they were using as its 'carrier' (adhering to it like dog hairs to a piece of Sellotape). Had Hahn and Strassmann been right, once they had eliminated the barium through a process known as fractional crystallization, they should have been left with radium. Yet, as Hahn wrote to Meitner on 19 December, such separation had proved impossible: 'It could still be an extremely strange coincidence. But we are coming closer to the frightful conclusion: our Ra isotopes do not act like Ra but like Ba . . . Perhaps you can come up with some sort of fantastic explanation. We know ourselves that [the uranium] *can't* actually burst apart into Ba.'[8] Meitner's response, two days later, underlined how surprising an outcome this appeared to be, while also demonstrating the open mind of the theoretician: 'At the moment the assumption of such a thorough-going breakup seems very difficult to me, but in nuclear physics we have experienced so many surprises, that one cannot unconditionally say: it is impossible.'[9]

It would not be long before the significance of Hahn's letter fully dawned on Meitner, although to get there she depended on a second collaborator, her nephew and fellow nuclear physicist Otto Frisch. It was while the pair were together for a Swedish Christmas that they made a breakthrough – reconciling an experimental result that seemed in direct contradiction with prevailing atomic theory. As Frisch later wrote: 'Gradually the idea took shape that this was no chipping or cracking of the nucleus but rather a process to be explained by [Niels] Bohr's idea that the nucleus is like a liquid drop; such a drop might elongate and divide itself.'[10] This was a bold theoretical leap, one made – as all great physics discoveries should be – while they were sitting on a tree trunk, pausing during a walk taken through the snow. Hitherto, physicists had allowed that firing neutrons at a nucleus might be able to dislodge a few protons and neutrons from it at most. By contrast, it seemed completely impractical that such a puny projectile could have so great an impact that it would break open and divide the entire nucleus. (Consider that uranium itself contains between 140 and 146 neutrons, and during the nuclear fission process absorbs just one.) It would have been a bit like assuming that one person diving into a swimming pool could completely empty it of water.

It was the liquid-droplet model of Niels Bohr, the celebrated Danish physicist who had first theorized that electrons orbit the nucleus of an atom, that prompted the realization. It allowed Meitner and Frisch to posit that the uranium nucleus was not actually a robust solid core, but something like a wobbling fat raindrop, needing surprisingly little encouragement to break apart. (As well as being visually useful, this image helped them to understand that the surface tension holding uranium's particles together was less than they had presumed.) As they sat scribbling calculations, aunt and nephew quickly started to piece the puzzle together. If this was indeed a wholesale division of the uranium atom, it would require a huge energy source to push the pieces apart; yet that could be satisfactorily explained by the smaller mass of the resulting products, and the transfer of that 'lost' mass to energy, in accordance with the most famous theory of Meitner's old friend Albert Einstein: $E=MC^2$ (which in simple terms defines the interchangeability of mass and energy). The idea that the nucleus had been cleaved in two, its fragments being forced apart from each other, helped to explain something they had all initially thought impossible: that the product of the reaction could be barium, an element with a substantially lower relative atomic mass.

Moreover, by accepting that barium and its fifty-six protons were one product of the divided uranium (which contains ninety-two), it followed that the other fragment must be an element with thirty-six, which they knew to be krypton. All of a sudden, an experiment that had led everyone involved to scratch their heads in disbelief was starting to make sense. Meitner and Frisch had, with stunning speed, assembled almost all the necessary clues. What they had concluded was both extraordinarily radical and intuitively sensible. It was a theory that clicked into place as if it had always been meant to be there. When Frisch shared the fission theory with Bohr just over a week later, he reported back to his aunt in a letter: 'The conversation lasted only five minutes as Bohr agreed with us immediately about everything. He just couldn't imagine why he hadn't thought of this before.'[11]

Even initially sceptical observers were quickly won over. Back in California, Luis Alvarez (he of the unfinished haircut) replicated the fission experiment in the Berkeley laboratory, and showed the results to Robert Oppenheimer. 'In less than fifteen minutes he not only agreed that the reaction was authentic but also speculated that in the process extra neutrons would boil off that could be used to split more uranium atoms and thereby generate power or make bombs.' Oppenheimer was correctly predicting the existence of a chain reaction, the root of how nuclear fission's discovery would be turned into a source of such huge energy, subsequently harnessed for two of the most significant inventions of the century: the atom bomb and the nuclear reactor.[12]

As the news of nuclear fission spread around the world at the beginning of 1939, physicists quickly turned their attention to the question of a chain reaction. As would subsequently be proved, the 'free' neutrons from fission (those not absorbed into either the barium or krypton by-products) can go on to generate further fission reactions: in one typical chain of uranium fission, the first reaction leads on to another three, and those produce another nine, with the spiral increasing exponentially from there.

A chain reaction was happening among scientists as well. The discovery of fission would set off a cascade of further collaborations as researchers scrambled to explore its implications. Like the barium and krypton products of the reaction, new partnerships were formed in the process. One of the most important was in America, between two European scientists: Leo Szilard and Enrico Fermi. It was their work that proved the existence of the chain reaction, and thereby paved the way for the invention of both the atomic bomb and the nuclear reactor.

It was a partnership, which is not to say it was plain sailing. The relationship between the two epitomized how scientific achievement can be the product of very different people with contrasting ways of working, as well as reinforcing how differences only really come to light when you start working closely with someone: learning about their quirks, how to make the most of them and manage around them, is one of the glories and the frustrations of teamwork. (Yes, all your standard office and personality politics goes on in the lab: maybe even more than outside it!)

Herbert Anderson, a PhD student who was Fermi's long-time collaborator and supported the pair's work on fission, recalled quite how different they were. 'Fermi's idea of doing an experiment was that everyone worked ... However Szilard thought he ought to spend his time thinking. He didn't want to stuff uranium in cans and he didn't want to stay up half the night measuring the [experiment].' For these tasks the cerebral Hungarian had hired an assistant of his own, who worked efficiently but whose presence irked Fermi. It was, Anderson remembered, the 'first and also the last experiment in which Fermi and Szilard collaborated [directly]. After that a mutually satisfactory arrangement developed in which Fermi and his associates did the experiments while Szilard worked hard behind the scenes to make them possible.'[13] This background support was far from incidental: Szilard helped to secure a government grant that enabled the work to continue, a negotiation that began with a letter which he wrote to President Roosevelt, signed by Albert Einstein, outlining the possibilities of this new discovery.[14] This all goes to show that collaboration doesn't have to mean poring over the test tube together: the people who do work behind the scenes can be just as crucial to progress within the lab as those who wield the instruments.

Szilard and Fermi may have been very different kinds of scientists, but they achieved a perfect compromise which harnessed the skills of both towards the shared goal. Their collaboration was a melding of two very different ways of thinking and working. 'Fermi was the careful scientist who planned every step, every detail,' remembered Victor Weisskopf, a physicist who went on to play a major role in the Manhattan Project that developed the atom bomb. 'Completely opposite to [Szilard]. Leo was averse to any kind of planning. He had the flashy kind of ideas.'[15] One such bright idea was to use graphite as the medium for the nuclear fission reaction. Counter-intuitively, this helps to maintain the chain reaction by slowing down the movement of the neutrons. It was Szilard who not only obtained the $6,000 that paid for large quantities of graphite to be used in early experiments, but who also worked out that it would need to be a higher grade of material (free from the boron impurities that negated the intended effect), and prevailed on the manufacturers to produce to his specifications.[16] By obtaining this bespoke graphite and gradually honing their model, the pair eventually succeeded in their goal: the first artificial nuclear reactor capable of reaching 'criticality' (the point at which fission becomes self-sustaining).

The reactor they built was the prototype of the power plants that provide around 10 per cent of the modern world's electricity supply. But you might not have guessed that looking at the strange contraption which 'appeared to be a crude pile of black bricks and wooden timbers', as Fermi himself described it. This was Chicago Pile-1, a 20-foot construction successfully tested on a derelict squash court at the University of Chicago in December 1942.[17] It achieved criticality under Fermi's watchful eye, measured with Geiger counters that had been named after characters from *Winnie-the-Pooh*, which the eminent Italian physicist had been using to help improve his English.[18]

This triumphant experiment was the culmination of many achievements made over multiple decades of research: a collaborative effort that was the sum of numerous collaborations and teams within teams, from Hahn and Meitner to Fermi and Szilard and many more besides. Like all culminations in science, it was not really the end of anything, so much as the beginning of a whole new set of research problems and potential applications as the world entered an era in which nuclear power would loom large and cause many more chain reactions of research and development. This is the essence of science: progress beckoned by discoveries that were built by many hands and inspired by numerous minds.

Yet even in the context of this bigger picture, it is impossible not to be attracted to the stories of the partnerships and small teams that drove different parts of this research effort forward: free neutrons that helped to make this explosion of discovery possible. They show that in science, like in so many parts of life, often it is difference that makes us stronger – aggregating complementary talents, contrasting ways of thinking, and different life experiences. However different two people may be, when they are both bringing something useful to the table, the basis for a strong and effective partnership is there. Even the most brilliant scientists tend to enjoy having other people around. And if it's good enough for them, then it should be for us. When some of the finest scientific minds who ever lived have acknowledged that they needed the help of their collaborators, research assistants and peers to make progress, it's a pretty good indicator that nothing important ever happens without teamwork.

The lesson from science is not just that you need other people, but that their skills should complement your own. A start-up that has one founder with amazing technical skills would probably benefit from another who can focus on managing the actual business with its accounts, people, customers and other issues that can't be solved through lines of code. A restaurant only works because it brings together people who prefer to work behind the closed doors of a kitchen and others who like to interact with people at front-of-house. These contrasts and differences are often the foundation for success. Collaboration is sometimes used as a warm and fluffy word, but in reality it is about knitting together different skills, personalities and outlooks on life. A little bit of friction can be just what you need, whereas having everyone agree all the time might be the worst possible thing.

I know that well from my work as a bioinformatician, using large data sets to try and work out how systems like proteins really tick. In this, there is a constant debate between biologists and software engineers about how to build models and slice and dice data. The software people always tend to pull for the most streamlined algorithm that they can promise will spit out neat and tidy answers. Whereas the biologists will point out that reality isn't quite so simple, and we need the best, most representative data to show a system in all its glorious complexity. Do you want the best-performing model or one ingesting the data that best reflects reality? The answer always lies somewhere in the middle, and it usually takes this tug-of-war between two different disciplines to find it: a perfect example of collaboration from a starting point of disagreement.

The need to collaborate is not just something for work, but a mentality to apply more widely outside your place of work.

There's not much in your life that won't benefit from a collaborative approach, whether finding a friend to put the world to rights after a break-up or getting someone else to review your job application (and even suggest going for a bigger or more ambitious role than you had considered yourself). If in doubt, to ask for help is not a bad mantra – and if you feel shy or self-conscious about doing so, then remember that science is giving you permission. There's a reason why the vast majority of scientific papers carry the names of multiple authors; however specialized an area of research, it is almost never a wholly individual pursuit. Science is less about colonizing and tackling a topic solo than it is about unearthing fertile ground for more discovery. Almost always, collaboration is the oil that greases its wheels, and the catalyst that delivers progress.

Collaboration in science is not just about people working together within networks or as individuals. It is also about the coming together of different types of scientists, and indeed different kinds of science. That is more than chemist working with physicist, theoretician with data analyst, or software engineer with biochemist, although those might be very useful collaborations in the right context.

It's also about the soft skills and the specific capabilities that are needed, first to bring a research project into being and then to take it to a meaningful conclusion. Think of all the different steps involved in some of the case studies we have explored so far. You have to find a good idea and hypothesis to explore in the first place, based on the ability either to observe anomalies, poke holes in existing theories, or perceive new possibilities. You have to go and get the right kind of funding and institutional support for that idea (which itself can also cause many existential crises, so well done if you have managed it). Then there is what we might call the work itself: designing robust experiments, interpreting the data that falls out of them, being able to solve problems as they arise and decide whether you are headed down an interesting pathway or a blind alley. Then, when you have some results, there's the debate that will follow about what you have (or have not) discovered: is it interesting, expected, relevant, and worth pursuing further (and if so, how and in what direction)? Can I do this on my own with what I have, or will I need to upskill myself or recruit someone else to help? And none of that even scratches the surface of communication: taking your ideas and findings out into the world through well-written papers and succinct conference presentations, hoping to find traction and support.

That is, to put it lightly, quite a lot to expect of one person or even a small team, even for those people who really believe that 'you can do anything you set your mind to, you just need to believe in it hard enough' (a pet-hate phrase). In the real world, the requirements for great change lean on different parts of the brain and varying personality traits unlikely to be contained in a single person. The researcher who can effectively pitch for funding may not be the same one who can nit-pick every line of the data with the necessary rigour (not at the same time, anyway). The person who solved a knotty problem with a control experiment may not be the one best placed to pitch the result to the outside world. Why get people to do the jobs they are less than perfectly qualified for, when you could build a team of people, each of whom excels in a particular facet? Don't get me wrong, I am all for transferable skills and learning from each other, however, if one person is judged for not being the other, this is where things start to get sticky – where people are not accepted for who they are or what they were hired to be. Instead, we should celebrate that great teams are the product of difference: miniature ecosystems that require many kinds of contribution to thrive, software and hardware working in sync. And we should value the soft skills that are easy to dismiss as less import-ant than cast-iron brainpower, but which are crucial to the process of scientific discovery as we know it.

Many of the scientists I spoke to pointed to the importance of collaboration in this vein, drawing on people who bring dif-ferent skills and divergent perspectives to the table, in service of a shared goal. Dr Leland McInnes, creator of the data-science technique UMAP (to weirdos like me, basically a celebrity), put it like this: 'A lot of good things come from the interdisciplinary aspect. My colleague from the machine learning side . . . what he's extremely good at is talking to a whole lot of people and then going, "Ah, there's one core problem here that's very inter-esting, that all these people all have." . . . He can phrase it in much more cleanly mathematical terms than the analysts would. But by the time he explains the problem to me in translation, I see it differently because I'm coming from an entirely different background, and then I can explain that to my other math col-league who will have a completely different viewpoint again.'[19] A chain reaction of storytelling that enables otherwise frag-mented perspectives to coexist and find common ground.

I love this idea of research problems being handed around like a mysterious object from an archaeological dig, whose original purpose is not entirely clear. Some people's impulse is to stick it under a microscope and look for small clues, others will want to play around with it and see how it works, still more will try and classify it against all other known objects that appear to be simi-lar. Think of them like your friendship group, and how all the different personalities contribute differently: the one who's good at breaking the ice, the one who tends to solve people's problems, the person everyone listens to. All these different techniques and mindsets have their value, and they might all be needed to solve a given puzzle. But you probably won't find an answer unless you have the different perspectives all gathered in the same circle. You need detail lovers and big-picture wallowers, those whose first instinct is to try and build up a story and others who relish poking holes in someone else's theory. 'I feel like that's where a lot of our progress in solving these problems comes from,' McInnes suggests. 'Just passing the problem around through different lenses and different ways of looking at it. And often that's different people with different perspectives from a different background.'[20]

Having people who think differently from one another, who come from different educational or social backgrounds, and who have contrasting life experiences, is not just an aid to problem solving and lateral thinking. It is also fundamental to ensuring equity in what results from scientific research and its practical applications in the real world. While science at its best can be a flag bearer for what a truly diverse team can achieve, it also has an extremely chequered history of bias, one that stems from a workforce that historically has been overwhelmingly white, male, neurotypical and non-disabled. The rub here is that we are finally in a world that is starting to value difference, yet which still does not understand how to work with and embrace truly diverse groups of people. Representation is part of the solution, but does not in and of itself equate to inclusion. Professor Christopher J. Hernandez, of the Mechanical and Aerospace Engineering School at Cornell, has written about how his upbringing as a Mexican American shaped, and in some ways inhibited, his scientific career. 'As a young boy, I was keenly aware that if I wanted to succeed in school, I had to be sensitive to rules and authority . . . pulling a prank might have been seen as cute if a white kid did it, but the same activity would get me labeled as a troublemaker or lead to punishment.' That awareness of authority, he suggests, often led him to take a cautious approach in his work as a scientist, pursuing experiments in which he had a high level of confidence, and presenting his work in a guarded way, noticing how his own 'cautious and defensive' writing style contrasted with a colleague's 'bold, confident language' on a collaborative grant proposal. The more his career developed, the more he came to see that this was an attitude shared with many fellow scientists of colour. 'I realized . . . that marginalized groups face a dilemma in science. The cautious strategy they adopt to succeed can ultimately be a hindrance for the sake of belonging, keeping them from reaching the upper echelons of science. In my case it helped me become a solid, productive scientist doing useful work. But it also kept me from pushing the boundaries of scientific knowledge.'[21]

Science's failures on equity and inclusion do not just limit its ability to harness the full range of talent in the world. They can also lead to direct and sometimes deadly harm caused by unconscious gender, race or disability bias. Consider the pioneering work done by Caroline Criado-Perez to shed light on how women have been harmed by a world that is invariably designed by men for other men – meaning a man was 23 per cent more likely to survive CPR resuscitation, because people had only trained on dummies of flat male chests, and that women (in a 2011 study) were shown to be 47 per cent more likely to suffer serious injury in a car accident, because car manufacturers were not configuring their safety equipment for female bodies.[22] Talking about the bonds of sisterhood, can we just take a moment here to acknowledge that whoever designed the cycling seat was most definitely *not* a woman.

The same bias extends to race: researchers at the University of Michigan found that the efficacy of pulse oximeters – which measure oxygen saturation in the blood, and were a life-saving tool for those with serious cases of Covid-19 – was considerably worse for people of colour. Their study found that a black patient was three times as likely as a white one to record a 'safe' level on the oximeter when in fact their oxygen levels had fallen into dangerous territory that required medical treatment.[23] Ending racism is a human right but it is also critical to the progression of science. It is only shared humanity, and efforts from both sides, that will undo the knots from the past and build inclusive futures.

These are just a few examples of a problem that has its root in the failure to work with truly diverse teams. A team drawn from a narrow subset of society is inevitably less likely to consider how their research will impact every member of that society: to foresee problems or ensure ideas and devices are tested robustly. If somebody has not experienced what it is like to be discriminated against, their ability to represent those who regularly are is going to be that much weaker. (This is why I got antsy when I heard that Disney hired a non-Asian lady to direct the new *Mulan* [2020] film. Much as we need to restore the degraded land on our planet of monoculture agriculture, we need biodiversity in every part of life, particularly science – the fertile ingredients that enable succession. After having the honour of winning the Royal Society Science Book prize (2020), I realized how this held a flare of light to those who were in the dark about their experiences before.

Teamwork is rightly celebrated, in science and beyond, for its ability to deliver ideas and innovations that none of the same individuals would have achieved alone. But it also needs to be subjected to a more critical lens: scrutinized for what it does not achieve, and held to account for flawed groupings that risk producing equally flawed outcomes. Simply having more than one person involved does not make research a collaborative exercise; for that you need people who are coming from different starting points, much readier to challenge each other than to fall into easy groupthink, and able to present a point of view or personal experience that a colleague could not. Nor does simply recruiting what appears to be a diverse team necessarily solve the problem. You can have a group with all the 'right' people in it, but without a conducive environment you will never empower them to do their best work. People who are used to being marginalized are also veterans in erasing themselves and excusing their differences, rather than harnessing the power of their unique minds. Without leaders who actively encourage them to be their best and truest selves, they will not escape the limitations society has silently imposed and imprinted over the course of years. Such work, fittingly enough, is a collaborative effort in itself, one that must be driven by the very people who think that a) the problem doesn't exist, b) that it isn't theirs to solve, and c) that unity can be achieved through HR emails and quarterly pizza parties.

Embracing difference is fundamental to scientific success, but so too, as Leland McInnes argues, is creating sources of commonality within a diverse team. 'We often talk about trying to build a gradient among the different people in our group. There's somebody in one specialty [and] it's very hard for them to talk to somebody else, but as long as we can put enough people in a chain between the two of them, we can all talk. That opens the line of communication and then progress gets made. It's managing to patch together teams of people that have the requisite overlaps.'[24]

Building true difference into research teams is one important way to facilitate collaboration in science – bonus points if you don't all agree on a subject. But it is also achieved outside the individual lab and project team, when people go further afield to take inspiration from unlikely sources. Whatever stage you are at in a scientific inquiry, it is never too early or late to cast the net wide and seek inspiration from unusual places. 'Senior scientists often encourage students to go to talks, especially the ones that aren't [about] their topic,' the neuroscientist Professor Rogier Kievit told me. 'It's quite tempting, and I had this temptation too as a student, to say I will [only] go to the talks that are sufficiently close to my field . . . But you see across a period of months and years that it's often the talks that are much further removed from your own field that will lead to interesting new questions and ways of thinking.' Even if this kind of response is only sparked by one or two out of ten such forays, he suggests, that will be 'invaluable, because they allow you to think of your own field in a different way than the other people'.[25]

That is an approach that could benefit people working in any industry. Whatever your job may be, it is inevitable that it is in some way subject to common expectations and approaches. There is 'a way of doing things', and most of us have a tendency to stick to that most of the time. There's no harm in this, but it can become limiting if not augmented with new perspectives. Because we are at our most curious and least confident (and so least cynical) when stepping outside the familiar, there is an active benefit to seeking out insight from people who do things that bear little or no relation to our own work. Also, it can take the pressure off having to be fabulous all of the time (which isn't easy, you know), allowing you to just relax into being an observer. Much of it may be irrelevant but, as Professor Kievit argues, the pieces of gold dust you do pick up are not just likely to be novel, but to be a potential source of competitive advantage. Sometimes the most useful collaborations will stem from the least obvious sources.

Collaboration in science is also something that transcends the individual partnership or the work of any particular team. In the view of the nanotechnologist Professor Jeremy Baumberg, who leads a nanoscience institute at the University of Cambridge, science itself proceeds as a kind of collaboration between two basic modes: simplification and construction. Simplifiers, he argues, 'take something in the world and they try and dig inside it and they say how does that work? . . . How does consciousness happen?

How was the universe made? How do galaxies form? These are all what I would call simplifier questions.'[26] In other words, simplifiers take things apart and try to work out how the pieces fit together, a logical approach to making sense of the extraordinary systems that abound in nature, and to decoding the limitless questions they contain. While simplifiers are trying to unpack and demystify what exists in the world, developing the theories and principles that explain how things function, constructors take a different approach. They conduct research essentially through building: developing models (actual or theoretical) that go beyond what already exists, and then prodding and poking these to develop even more questions and lines of inquiry. If the philosophy of the simplifier is to explain how something works, the creed of the constructor, Professor Baumberg argues, is to ask questions like: ' "What happens if I do this to that?" or "How might this operate if I heat it?" ' The simplifier asks "how it works" while the constructor ponders, "How can we make it work differently?" '[27]

Hands up! I am definitely more of a simplifier and proudly so. I love delving into things, discovering everything I can about them, becoming obsessed with every new branch of the tree that I am climbing. It's why I poured years into doing a PhD in my first love of biochemistry. However, over time, the more deeply I read into the subject the more stale it felt. I knew there was more to science than this: data to be grabbed and new worlds to be built with it, *Minecraft* style. So after completing my doctorate, I applied myself just as obsessively to machine-learning research, brute forcing my way through a course to learn the programming language Python, and developing my career as a bioinformatician. Welcome to my first existential crisis, which at one particularly low point involved a near miss in me purchasing a 'high-tech mechanical keyboard' in a pilgrimage to embody the mind of a software engineer.

So there is a constructor side to me too, and I revel in being able to do both the exploration and the building. There are some challenges to this – I sometimes get branded as the more emotional side of technical, which despite my co-dependency on Python working, doesn't make for a textbook software engineer – but for me the combo is where it's at. And it shows that, much as I love Professor Baumberg's typology, the two don't have to exist independently of each other.

You can be both a fiend for scientific literature, and someone with an aching desire to create, program and own a new project. A theoretician who doesn't mind getting their hands dirty. These contrasting pieces, like any good collaborative relationship, can be mutually supportive. Building and doing can give you a fresh perspective on how you think, while reading and learning will obviously influence the way you build. Loving cookery books doesn't exclude getting in the kitchen to make up or tweak your own recipes. In this way you can almost collaborate with yourself, the different sides of you feeding one another. You aren't required to climb into one box or another and make it your home for life.

And even if on an individual level you feel more constructor or simplifier, in the grand scheme of things science needs them both working together. A world with only simplifiers in it might get nothing done, but one built purely by constructors would likely fall apart, lacking the solid underpinnings of what their counterparts have already worked out. 'Constructor science is based on initial simplifier discoveries and would simply not be possible unless those discoveries were robust in almost every way,' Professor Baumberg has written. 'Imagine the scientific landscape as a city of buildings. The architectural toolbox produced by the simplifiers includes cables, concretes, columns, struts, and beams. Constructors put these together in imaginative ways.' Yet they would not actually succeed in building anything without that toolbox that their colleagues have provided. 'The reason we don't have a cacophony of scientific rubble crashing to the ground is the sheer robustness of most of our knowledge.'[28]

This scientific *Sim City* is only possible because of the innate collaboration that happens on a large scale between the scientists who patiently tease out new discoveries, and those who then take that knowledge and expand it in bold new directions: the brick makers and the cathedral builders. Professor Baumberg gives the example of the particle accelerator, which in its primitive form helped to enable the discovery of nuclear fission in the 1930s, and today continues to decode the secrets of the atom through pioneering apparatuses such as the Large Hadron Collider. This, he suggests, is a prime example of the dance between simplification and construction that is ever-present in scientific progress. The immense electromagnetic fields which power the LHC could only be constructed based on the development of new materials that were the product of simplifier research.[29] In the same way, we can see that the development of the atom bomb and the nuclear power plant were constructor projects that rested upon the simplifier work of Hahn and Meitner, and their many colleagues and predecessors in nuclear physics. All the work to simplify our understanding of the atom, its structure and its propensity for reaction, helped to make possible the development of the first artificial nuclear reactor – which in turn rested on the constructor science of Fermi and Szilard, working out the right kind of graphite they needed to use, packing tin cans full of uranium oxide and stacking their nuclear 'pile' higher and higher. The birth of the atomic age, therefore, should be seen as a child of scientific collaboration on numerous levels: between the individuals who drove forward different areas of research, across the entire network of nuclear physics at a time when the field was rich with new discoveries, and between the different forms of science itself – the simplification and construction – that are both highly contrasting and ultimately interdependent.

As this shows, collaboration is something that comes in numerous forms and which contains endless possibilities – so much more than a group of people sitting in a room with a whiteboard and marker pens. Science shows us that the true benefits of collaborative working only come when you understand the full breadth of the field in which you are working, and all the different people in it, tackling problems in their own particular way. Failing to sniff out that ever-expanding knowledge is a bit like trying to navigate a dark room with a torch when you could just hit the light switch. Your opportunities to work with people go so much further than those with whom you share an office, or those you think may be most relevant to your field.

At the same time, it matters how you work with the people around you, who is invited to be part of that team, and most importantly who we put faith in and why. Effective collaboration isn't achieved by agreeing with each other the whole time. It should be about different kinds of people who bring contrasting perspectives and skills to the table – this isn't 'noise', but more likely the ingredients to challenge an assumption than to run with it. Science teaches us that friction between people is often what is needed to make the wheels of an experiment turn. The truly collaborative partner doesn't pat you on the back: they scrutinize your work and identify its flaws, and then help you to do better – like Lise Meitner telling Otto Hahn that he needed to test his 'radium' again, because it really didn't sound as if he'd identified the right element. True teamwork is made of critical friends like these, doling out tough love for the sake of truth. They share a common purpose but will often disagree about how best to achieve it. These disputes need to be managed carefully to avoid becoming acrimonious, but they are generally the sign of a healthy team in which the competition between ideas takes precedence over ego, hierarchy and groupthink. And they hold lessons for life in general: be willing to listen to the colleague who disagrees with you (in fact, listen to them twice as hard); ask yourself why it matters to your partner or house-mate that the dishwasher is stacked a certain way; and don't discount the relative or friend who criticizes one of your life choices just because it makes you feel uncomfortable. Science teaches us that we have to be open to feedback and aware of our own limitations. That is where other people, their contrasting skills and knowledge, and our ability to work with them, come in. And any scientist will tell you that healthy debate, and even vigorous argument, is something to be welcomed rather than feared. Despite our preconceptions, disagreement is often more an enabler of progress than a barrier. By contrast, it's when everyone is nodding quietly in agreement that you need to start worrying.

7.   Proof

How to seek (but not always find) answers

In science, we are used to hearing about old theories subsequently being proven wrong, or at least incomplete. There is an assumption that science works progressively, building upon and mostly improving what has gone before, as practitioners benefit from the growing mass of accumulated knowledge and improved tools of the trade – towers of knowledge that stack up ever higher, until the ideas on which scientists previously relied are out of sight. But sometimes the boot is on the other foot, and scientists succeed not in displacing old theories but validating them, exploring new implications of ideas that were first put forward long before they were born. Such is the long and winding path that science takes towards establishing the proof of any fundamental idea.

There is no more famous example of this than Einstein's theory of general relativity, one of the most debated and tested concepts in the history of physics. First published in 1915, it evolved the ideas he had already developed around special relativity – that the movement of any object through space and time is never absolute but always relative to something else, and that the constant factor is the speed of light, which nothing else can exceed. General relativity expanded this vision of space and time to include gravity, suggesting that its interactions with space and time were dynamic, causing them to warp and bend, as if made of memory foam (hence those pictures you may have seen of what looks like the earth sinking into a piece of graph paper).

The ramifications of the theory were extensive. It predicted all sorts of things that would subsequently be proven, from the behaviour of light passing through a gravitational field (it bends) to the existence of black holes, areas where a star has collapsed and the force of gravity is so strong that no light can get out. In the decades that followed the introduction of general relativity, a huge number of experiments were done to explore its implications, including some quite wacky ones. In 1971, two researchers flew twice around the world – once in each direction – with four huge atomic clocks as passengers.[1] This demonstrated, as Einstein's theory had predicted, that time is indeed experienced differently relative to the position of the observer, appearing to move microscopically faster with greater distance from earth's gravitational field. (A cool experiment, but just imagine trying to get a massive piece of lab kit through airport security these days, and reassuring them that it's all fine, it's for science. And you don't even want to know about the excess baggage fees.)

As time went on, at whatever speed you happened to be observing it, the experiments continued and the proofs kept rolling in. It wasn't until 2015, a full century after general relativity had first caused a stir, that its final component was validated. An experiment found the first proof of gravitational waves – released by high-energy cosmic events, like black holes merging or very dense stars colliding, and causing literal ripples across the fabric of spacetime. This was an implication of general relativity that Einstein had predicted, and one of potentially huge importance to physicists: the ability to measure these waves could transform our understanding of how the universe is expanding, and how it has done since the very beginning of time.[2]

The detection of gravitational waves in space – a full hundred years after they were first mooted – demonstrates the complicated road that must often be trodden in science between theory and proof. Thinking something exists, knowing it to be the case and demonstrating it beyond reasonable doubt are all different things. Each of them may be the stage in an important concept evolving from interesting hypothesis to viable theory and proven entity.

It's not as if scientists were dragging their heels on gravitational waves until 2015. Experiments and research went on throughout the intervening years, getting us gradually closer to closing the gap between theory and proof. One physicist claimed that he had recorded them in 1969, using giant aluminium cylinders as detectors, which he said functioned like tuning forks. The only problem was that no one else could replicate the result. A few years later, two astronomers observed a pair of neutron stars behaving in a way that strongly indicated the existence of gravitational waves – work for which they were later awarded the Nobel Prize. These experiments, among others, took scientists much closer to accepting that the phenomenon was real.[3] But observing the effects of something at work is not the same thing as actually detecting its presence. Scientists don't generally put up their feet and get out the sun loungers when something has been shown to *probably* be the case. Knowing that something is almost certainly present, even if it can't actually be shown, doesn't quite scratch the inquisitive itch to satisfaction. So the work continues until you have some actual evidence. (Which may not in itself be proof, or prove anything beyond the specific context in which you have observed it. Difficult, right?)

But we shouldn't despair about the search for proof being so long and arduous, and producing so many results that don't quite hit the mark. Because discoveries that fall short of proof are not failures, but one of the most important catalysts for progress. There are social dynamics in science like any other field: researchers can back winners just as much as glory hunters align themselves to a successful football team. The prospect of an important breakthrough can increase interest and, critically, funding in a particular area of research.

A second catalyst is technology. One of the main differences between any two generations of scientists will be the tools at their disposal. You no longer have to drag a massive clock onto an easyJet flight to tell the time at different points in the earth's orbit. Today's astrophysicists and cosmologists have not only vastly improved telescopic and imaging technology, but the power of modern computing to crunch the numbers that result. The constant march of technology is something scientists depend on, but which they sometimes find themselves waiting around for (in the way that it took James Cameron thirteen years to make the sequel to *Avatar*, in part because he wanted better cameras for underwater motion capture).

The LIGO (Laser Interferometer Gravitational-Wave Observatory – quite a mouthful!) project that discovered gravitational waves in 2015 was just one example of when researchers imagined tools that would not be available to them until many years later. Two of its originators, Dr Kip Thorne and Dr Rainer Weiss, first met forty years before the experiment that saw their work bear fruit. The equipment that ultimately delivered the momentous result had been conceived even earlier. Weiss first wrote a paper about the theoretical design of LIGO in 1972. Construction began in 1994 on the apparatus, which would finally consist of two steel-and-concrete arms 2.5 *miles* long, supporting mirrors hanging from glass threads, monitored by lasers capable of detecting a movement significantly less than the dimension of a subatomic particle. It did not come online until the early 2000s, and in the 2010s was redesigned to improve its sensitivity.[4] This decades-long process to design an experiment that satisfied a century-old theory gives some indication of the time scale on which science sometimes operates. Proving one theory may first require proving another, about the experiment you wish to conduct and the method you intend to use. As this suggests, big visions are as much about perseverance as inspiration. Another reminder from science that patience is one of the best tools available to us, whatever we are trying to achieve.

The LIGO investigation of gravitational waves reveals the complexities and dependencies that go into achieving what we so sweepingly call proof (a word I have arguably been misusing up to this point, as I will explain). And it indicates how scientific knowledge and progress rest on a fragile relationship between the theories scientists develop, the evidence they base them upon, and the degree of proof they can advance in their favour. All are tricky, fluid and hotly debated concepts. Theories may or may not be provable (and they can still be useful even if they haven't been proven). Evidence may or may not be complete, relevant, replicable or trustworthy. Proof may or may not be convincing or comprehensive. It may, depending on your philosophy of science, not even be possible.

Nor is it always necessary for an idea to be useful. Proof is a sexy idea in science: the belief that a hypothesis will eventually be either proven or disproven, validating months or years of work. But the scientific pudding can actually be very tasty even when no proof has been involved in its baking. Much of theoretical physics is based on concepts that lack any empirical evidence, but which have proven to be hugely important in shaping our understanding of the universe (recall how we are still searching for dark matter and dark energy). Conversely, scientists can find proof for all sorts of things that are essentially useless when it comes to advancing our collective knowledge or intelligence. The ability to validate and prove a scientific idea isn't necessarily the slam dunk that it first seems.

The search for proof is one of the most important and illuminating aspects of the scientific profession, shedding light both on how inquiries can advance and how they may go wrong. The gaps between theory, evidence and proof are where we can see some of the best scientific work – defined by an impatient but also long-term approach – and some of the worst pitfalls, where the desire to find proof leads the evidence to be twisted, consciously or unconsciously. It is when we see researchers looking for something they are convinced is out there, challenging a theory that has been accepted for decades, or searching for the missing piece that would complete a puzzle they have been working on for years, that we can learn some of science's most important lessons.

So far we've been proceeding on the basis that proof in science is something that exists, which is desirable, and should be the ultimate goal of all scientific research. Forgive me, for it's quite a bit more complicated than that. The very idea that something can be proven in science invites its own challenges. Has it been proven universally? Proven beyond reasonable doubt? Proven across all dimensions of space and time we haven't discovered or conceived of, as well as those we have?

One problem with the idea of proof is that it suggests finality in a process that, by its nature, has no end. In science we have observations, which inform hypotheses, which may develop into theories, and harden into principles or laws. Yet none of these things are ever certain: all are subject to being disproven, superseded, or at the very least shown to be incomplete. There will always be people who come along and look at a familiar problem through a different lens, point out inconsistencies that have gone unnoticed, or who ask pertinent new questions. When is proof complete? When can we shut the curtains on an experiment and declare that we are finally done? The issue of what has actually been proven, and to whose satisfaction, is never entirely settled: as we will see, even concepts with the words 'universal' and 'law' in them may turn out, in the fullness of time, to be neither of those things.

The very basis of how science is done, with lots of people prodding and poking ideas, thinking critically and sceptically, ensures the instability of its foundations and the vulnerability of anything that gets labelled as proof. Scientists generally hold most theories and laws to be somehow guilty – whether of being incomplete, incompatible with other scientific principles, or insufficient for the scale of the problem they seek to govern. It is much easier to disprove a theory by showing it falls down in one particular aspect or context than to prove that a particular law holds true in every actual or theoretical situation it may encounter. As such, a lot of the progress made by scientists begins not with the creation of something but with destruction – if not quite tearing down existing theories, then at least pointing out the gaps in them, and grabbing hold of the unanswered questions that fall out. We find it easier to poke holes first, but doing so is not necessarily vandalism: it can also open up a new way of seeing.

The question of whether science should be creative or destructive, primarily focused on building Jenga towers or removing blocks from them, was a major debate in the field during the mid-twentieth century. On one side was the philosopher Karl Popper and his doctrine of falsification: that a scientific theory was only valid if it was capable of being refuted, and that scientists should be working not to prove their ideas but to disprove them: 'Every genuine *test* of a theory is an attempt to falsify it, or refute it . . . Confirming evidence should not count *except when it is the result of a genuine test of the theory*, and this means that it can be presented as a serious but unsuccessful attempt to falsify the theory.' By contrast, he believed: 'It is easy to obtain confirmations, or verifications, for nearly every theory – if we look for confirmations.'[5] In other words, if you are desperate to be proven right about something, you can usually find some kind of evidence to look as if you are. But the true scientist,

Popper argued, attempted to kick over their own sandcastles, not dig more moats around them. They actually *wanted* to prove themselves wrong.

In the opposing camp was Popper's contemporary Thomas Kuhn, who suggested that most scientists, most of the time, were using established tools and theories in the practice of what he called 'normal science'. Only when these had repeatedly shown themselves to be unsuitable would attempts be made to rewrite the rulebook, breaking with one 'paradigm' and forging another. '[When] the profession can no longer evade anomalies that subvert the existing tradition of scientific practice – then begin the extraordinary investigations that lead the profession at last to a new set of commitments, a new basis for the practice of science.'[6] On this basis, major breakthroughs arise in science through the gradual process of realizing that the old techniques and ideas have become redundant, rather than turning up for work each day desperate to prove that they are wrong.

Whichever of these views you happen to prefer, they take different paths to the same point: it is dissatisfaction with the scientific consensus that ultimately leads researchers to advance their work and break new ground. A very human quality of not stopping until everything lines up or feels right. We all feel this to some extent, whether we are a Popper (dissatisfied by default) or a Kuhn (following the recipe until it lets us down and we have to find a new one). I am guilty of this myself. With a large data set I often can't help skipping past the obvious similarities to focus on the minuscule differences, looking for details that might not seem important, but which could contain unturned pebbles that are concealing something significant. With data like this it's often easy to see the big and important conclusions, but I am never happy until I have spent time hunting out minor quibbles and inconsistencies, just in case.

Either way, we can see that scientists are as much motivated by doubt as they are fired up by the promise of discovery. The observation or experimental result that *doesn't* make sense is often so much more interesting than the one that was expected. It is the anomaly that starts to cast light on the flaws in a particular theory, opening the eyes of researchers who can quickly start to flesh out an emerging hypothesis, or spot fresh lines of inquiry.

Such an anomaly might be tiny: an almost imperceptible nuance or apparent error in a data set. Or it could be absolutely bloody massive, the size of a literal planet. Let's go back to your friend and mine, the theory of relativity. When Einstein was putting the finishing touches to his masterpiece in 1915, he set three tests to establish its veracity, the last of which was its ability to predict the orbit of the planet Mercury. This was a well-established sore point in the theory that had long dominated Einstein's field of study: Newton's law of universal gravitation. This winningly simple and highly successful formula holds that all objects exert gravity on each other, and that the gravitational force between two objects can be calculated as long as you understand their masses and the distance separating them. This became the proverbial hammer with which to pin down new discoveries. It envisaged a fixed world in which everything could be explained by relative mass and position, in contrast to Einstein's highly dynamic, more nuanced concept of general relativity, in which objects literally bend space and time as they move through them.

For more than two centuries, Newton's theory was pre-eminent, helping to explain and open up the study of everything from the planets downwards (and it remains a good, if incomplete, approximation for our understanding of gravity today). But one thing it could not explain was the orbit of Mercury – the planet closest to the sun. Unlike the other planets, measurements of its perihelion (the point in its orbital cycle when it flies nearest the sun) were inconsistent. There was a tiny but unambiguous discrepancy in these measurements that universal gravitation suggested should not have existed. As it passed close to the sun, Mercury was moving just a tiny bit faster than Newtonian gravity said it ought to have been. Einstein did not discover this anomaly, but he was the one to resolve it, with general relativity accurately explaining the planet's movements in a way the forerunner theory had not.

By stating that Einstein's theory for gravity (published in 1915) superseded Newton's (published in 1687), we are skipping to the end of a rather long story, which has a lot to tell us about how scientists seek proof and sometimes go astray in the process. During the second half of the nineteenth century, there was an extended and fascinating period in which scientists tried to make sense of the curiosity of Mercury while working within the 'paradigm' of Newton's laws. As Thomas Kuhn argued, scientists do not necessarily abandon a theory just because of a few anomalies, even if they are big ones. The first scientific instinct is generally not to throw everything out of the window but to justify the rogue finding, to explain how it can all make sense within the bounds of a particular theory or law. Often that means researchers trying to locate something they may be missing: a force or variable within a system that, if identified, would account for the anomaly.

This is what happened as astronomers and physicists tried to get their heads around Mercury's wonky perihelion in the context of Newtonian gravity. The obvious deduction was that they were missing something: an object whose gravitational pull would explain why Mercury and the sun were not interacting the way the universal law said they should. So they set out looking for it, buoyed by the fact that similar logic had led to the identification of Neptune in 1849, after discrepancies were recorded in the orbit of Uranus.

The French astronomer who had predicted that discovery, Urbain Le Verrier, believed that the same would hold true with Mercury, and that a ninth planet orbiting between it and the sun would provide the explanation for that pesky orbit. After he had published this theory, astronomers got busy with their telescopes, eager to be the first to track it down. The putative planet's trajectory so close to the sun meant it would be incredibly difficult to observe, but before long a sighting had been made, by a French doctor and amateur stargazer, Edmond Lescarbault, on 26 March 1859. With his telescope, he spotted a tiny dark object moving over the surface of the sun, one that was visible on this trajectory for just over an hour and seventeen minutes. Only months later, reading an article by Le Verrier, did he realize what this might have been. In December he wrote to the famous astronomer and received something far more than a reply.[7]

So excited was Le Verrier by the news of this potential sighting that, although it was New Year's Eve, he got straight on a train to visit Lescarbault in his Brittany village and hear more. He was unperturbed by the fact he had a party to attend, or that a 12-mile walk awaited him at the other end. It's not hard to imagine how the hobbyist must have felt as the world expert turned up without warning, presumably the worse for wear from his hike, and demanding to inspect his instruments and question every detail of his work. Yet remarkably – for Lescarbault was working in some cases with home-made equipment, had recorded his findings on a piece of wood for want of paper, and had had to interrupt his observation of the mysterious black spot when a patient arrived to see him – Le Verrier declared himself satisfied. In January he announced that the new planet had been discovered, pairing Lescarbault's observations with his own calculations about its orbit and position. By February it had earned the name that would make it famous: Vulcan, after the Roman God of fire, volcanoes and the forge – appropriate enough for what was thought to be the planet that flew closest to the sun.[8]

The announcement that a new planet was not just theorized to exist but had in fact – it seemed – been observed raised excitement to a new level. Vulcan's position so close to the sun meant that a full solar eclipse offered the best prospect of sighting and studying it in detail. Whenever one occurred, stargazers would ready themselves in the hope that they would be the one to validate the theory of the new planet. In 1878, Thomas Edison even joined a group of astronomers to study an eclipse covering North America, one of whom believed that he too had sighted Vulcan, causing another media stir. Yet like Lescarbault's apparent sighting almost twenty years earlier, this one went unrepeated. The astronomer who had made it, James Craig Watson, died just two years later at the age of forty-two, as he was overseeing the construction of a massive telescope designed to let him observe Vulcan without the benefit of an eclipse.[9] Le Verrier himself had died three years earlier, but the idea of Vulcan persisted. Only with Einstein and the arrival of general relativity in 1915 did the hunt for it really abate. He had created, in Kuhn's terminology, a new paradigm for gravity; there was no longer any need for creative interpretations to make Mercury play nicely with Newton's law. The over-promised and under-delivered planet Vulcan didn't actually exist in the first place! Showing the risks of squeezing what you see into a theory held as default rather than it being a basis for defining proof.

The pursuit of Vulcan, the planet that never was, gives some insight into the dance between theory, evidence and proof that is such an intrinsic part of science. Vulcan was a theoretical object to satisfy another theory by eliminating its most obvious anomaly. The only tangible evidence underpinning its presumed existence was Mercury's wonky orbit. Everything else was purely theoretical or circumstantial: the belief in Newtonian gravity, a handful of sightings that could have been observing all sorts of objects if they were anything at all, and the notion that, because one anomalous orbit had been explained by the discovery of a new planet, the same would hold true in this case.

It is not that Le Verrier was a bad scientist. In some ways his work was extraordinary, with his measurement of Mercury's perihelion being extremely close to the one recorded by modern technology, a remarkable achievement given the tools at his disposal. But he was a product of his time, the scientific consensus as it then stood and his own career history. As someone who prayed at the altar of Newtonian gravity and was in the business of adding strings to the solar system's bow, the theory of a new planet made sense – in fact it made the most sense. With hindsight we can see that Le Verrier was erring when he took that 12-mile walk to see a country doctor who did astronomy for a hobby. But to him, and many (though, it must be said, not all) of his contemporaries, the concept of Vulcan was more plausible than anything else that explained Mercury's behaviour. Throw in too that scientists get excited by the idea of big new discoveries, and you have the ingredients for a wild-goose chase across space and time. As this shows, the search for proof can become an end in its own right, a siren call which leads even good scientists astray. Their desire to find something becomes so great that they will themselves to see what simply isn't there. A long walk for a ham sandwich.

This all helps to underline what a fickle concept proof is in science. For more than half a century, scientists studying the solar system tried to prevent an important theory from being disproven by dreaming up a theoretical object that itself would never be proven, because it had never existed. The upshot was that a new theory would come along that both superseded

Newtonian gravity and rendered the hunt for Vulcan redundant. That is the drumbeat to which science marches: slowly, sceptically, sometimes reluctant to surrender an established consensus, but always eager to explore the early glimmer of something new and revealing. At the same time, it exists in a constant state of flux, in which established ideas may provide a solid scaffold, but never a cathedral that will stand for a thousand years. 'You can always prove any definite theory wrong,' the great physicist Richard Feynman said in a lecture in 1964. 'Notice, however, we never prove it right.' For all the time that Newton's concept of gravity had gone unchallenged, he argued, 'the theory had been failed to be proved wrong, and could be taken to be temporarily right. But it can never be proved right, because tomorrow's experiment may succeed in proving what you thought was right, wrong.'[10] There's a reason that established scientific concepts are often described as principles, theories or laws but never as 'proofs'. Principles can change, theories can prove mistaken, and laws can be changed or updated. But proof is absolute and final. And science, you will agree with me by now, is never either of those things.

So if we accept that nothing is ever certain, and nothing can be proven in a way that might not be one day overturned, then what are scientists left with? The answer – which, ironically, I am pretty certain about – is quite a lot, actually. That's because the fallibility of scientific knowledge is not a fatal flaw but a secret weapon. That nothing is ever settled beyond reasonable doubt is just a different way of saying that everything is open to being explored – subjected to new tests, observed through different lenses and cast in a fresh light. This simple reality provides the impetus for pretty much all scientific research: there is still so much interesting work to do. It's when things seem settled that you actually need to worry.

It should be recognized that the lack of conclusive proof does not prevent the theories and laws scientists develop from being extremely useful. After all, the entirety of modern life is based around technologies that were developed based on scientific principles. Theory, and the necessarily incomplete evidence base on which it rests, can do just fine without conclusive proof, and the search for conclusive proof is a bottomless pit which has sucked in many a perfectionist. That's right, you actually don't need to start when you are 100 per cent ready or justify action from boundless preparation, otherwise you would be waiting for ever. Helpful advice for those of us with a tendency to procrastinate and use 'research' as the excuse for not getting started on a new project. You can never be sure about anything before trying it, so you're better off just getting on with it and learning as you go. This is especially important if you are marginalized: when you don't 'fit' the mould of an industry or community, you can spend for ever trying to justify your presence, and thinking that you need to act in a certain way. The 'proof' that someone like you can succeed in this environment may not exist. It might feel as though you are constantly being questioned. But if you want to succeed, you cannot always wait around for invitations. There's no preparation that can beat simply showing up.

In their own way, the emerging scientific theories can also be a touch haphazard: more a case of suck-it-and-see than turning up to collect the Nobel Prize on day one. When general relativity arrived, it did not sweep away Newtonian gravity and declare it heresy, never to be spoken of again. It simply offered a more complete, more universal explanation for the phenomenon, one that has inspired countless extra dimensions of research, just as Newton did in the seventeenth century. The universal law of gravitation is still a good rule of thumb, one that modern equipment has proven accurate to a surprising degree of detail.[11] In turn, more than a century after it was announced, Einstein's most famous theory itself continues to be scrutinized and, as physicists delve into additional dimensions of spacetime from the four that he outlined, it will itself almost certainly suffer the same fate.

As old theories get superseded, the search for new ones continues, often leading to the scientific white whale of one theory to rule them all. The ultimate, big cheese of theories. As the theoretical physicist Marcelo Gleiser has argued, there is a broad tendency in science to go beyond the basic discipline of working to unify theory and evidence – 'connecting as few principles with as many natural phenomena as possible' – and instead seek something more: 'the *Uber*-unification, the Theory of Everything, the arch-reductionist notion that all forces of nature are merely manifestations of a single force'.[12] As he describes, such attempts have been a pursuit of some of history's most famous thinkers, from Plato to Einstein (who had unsolved equations for his Theory of Everything written on his blackboard when he died) and Stephen Hawking. And they have been brought into question by ideas such as Gödel's first incompleteness theorem, which argues that any mathematical system that applies universally (even addition and subtraction) must produce results that cannot be proven by that system even if they are true. In other words, unprovable truths exist in a system by its design. I often like to think of this as a parent describing the wise ways of adulthood to a child who doesn't yet know their worth; only through time (as the system changes) do they realize that their parents – dare I say it – were right. Despite continuing efforts, the sceptics have been winning on this one and unification remains far off: physicists are still struggling to reconcile contradictions in the theories on which modern understanding of their field is based, such as the conflict between general relativity and quantum theory. And that is before we get into the realm of what has not yet been explored, discovered or even conceived of. What unified theory could ever fully capture a universe that we are so far from truly understanding? Gleiser argues that the search is futile: 'You can conclude that what we do best is construct approximate models of how nature works and that the symmetries we find are only descriptions of what really goes on . . . However, let's not confuse our descriptions and models with reality.'[13]

Proof is a funny old thing for scientists, because it's both the thing we seek – to the limited, non-absolute extent that it exists – but also the promise that risks leading us astray. Scientists are compelled to dig their chisels into small gaps in existing ideas and see if anything gives way; and they cannot help themselves from pursuing the glimmer of a new idea, even if it is a bold and perhaps unlikely one. Both of these instincts are necessary ones, but there are pitfalls. At the risk of having my membership to several secret societies cancelled, it should be acknowledged that any scientific theory is essentially a fantasy: an idealized version of how a particular system works, in which a scientist arranges the pieces as a novelist does characters and plotlines. There may be convincing evidence for large parts of this picture, and the fantasy may prove to be a useful way of achieving clarity, but it remains no more and no less than a plausible illusion. And as we saw with the example of Vulcan, such illusions can leave scientists chasing the ghost of something they believe to exist, but will never be found. The impulse to find proof drives on the work of science, but it can also lead researchers down blind alleys and incentivize them to lose perspective as they pursue a proof that they have become determined to realize. This is a warning we would all do well to heed: every one of us can be distracted by shiny objects that appear to hold the answers to our problems. But a big Insta following isn't proof of your popularity any more than a fast car is of a man's, er, masculinity. Sometimes the things we seek do more to lead us astray than towards meaningful goals. And just in the way that a theory without proof can be a wonderful and useful thing, so too can a life be happy and fulfilling without many of the status symbols we have been taught to associate with success. One of the cautionary tales of science is not to let your focus on something become so myopic that it makes you forget about what really matters.

It's also important to recognize that the way we seek proof is changing as technology transforms our ability to process data, produce models and build complex simulations. Modern software can often achieve in moments what would have taken researchers many months or years using traditional methods. One notable example is AlphaFold, a program that uses AI to predict and model the entire 3D structure of a protein based on a single amino-acid sequence. When DeepMind, the company behind the software, announced in 2022 that it had cracked the code of over 200 million proteins – nearly all of those known to exist – it caused a huge stir in the scientific world.[14] This was an astonishing achievement, not dissimilar to clicking your fingers and watching an entire house be constructed before your eyes in seconds. And it has huge implications, particularly in the field of drug discovery, where it could massively speed up the search for vaccines and treatments for deadly disease. Proteins really are the facilitators of the natural world in almost every form – the more we know about them, the better we can understand and design helpful interventions, whether to defend the body against hostile forces or solve environmental issues such as how to decompose plastics. AlphaFold will transform how scientists in many fields are able to work, and is just one example of the advances being made through the use of AI. But it also has limitations. It's a computer programme, crunching data and calculating probabilities to produce indicative models that may not be entirely accurate (and which it recognizes by grading regions of its models on confidence intervals from 0 to 100, where anything from 50 to 70 is 'low confidence and should be treated with caution' and anything below 50 'should not be interpreted'). Professor Janet Thornton, director emeritus at the European Bioinformatics Institute, has described it as 'an infinitely clever cut and paste algorithm'.[15]

This is not to play down the importance of AlphaFold or the significance of AI across science in general, but simply to recognize what it is and does. It is modelling and prediction, not (as some of the media coverage implied) determination or proof. It gets us a long way very quickly, but is an approximation of the more complete picture a biochemist creates when given the necessary time (and funding) to do detailed experiments. It's a good enough model to work with, but it isn't evidence that you could take to court – just like how a rose-tinted social media post from a friend doesn't substitute for conversations over dinner, or being there in hard times.

Does this matter? Well, yes and no. It's important that we don't mistake computer models for evidential proof, but also the case that the existence of all this computer technology may change how we think about proof. Why try to conclusively settle the matter when you can just get the machine to spit out another set of results, and then another, until you find something that allows you to make progress?

This brings into focus the many problems with the idea of proof in science. It's a word that suggests conclusive, universal and undisputed truth – which are all heavier burdens than any scientific principle can bear. It implies that the journey of scientific discovery is one that goes neatly from beginning to end, rather than being a rambling series of missteps, detours and dead ends, of which the only real conclusion is the passing of the baton on to a new generation. There is always a painful and necessary preamble to an overnight success. The limitations of scientific inquiry must be recognized: we can demonstrate certain things to be true in certain contexts, but never achieve a proof that shows something will remain the case in every possible context with which a theory might intersect. Proof isn't singular: more a spider's web that is strong in some areas and fraying in others.

This is the agonizing reality of science – the harder you look at a particular problem or principle, the more you realize how complicated it is in its totality, and how unrealistic the idea that anything can be scientifically proven really is. It's for this reason that you will get a much more confident answer about how gravity works from a primary-school class (who know it's the force that makes things fall down) than from a seminar of theoretical physicists (who still wonder about its quantum properties, its subatomic structure, and frankly what even causes it to start with). Yet the elusive, illusory nature of proof isn't something to make a scientist down tools and give up hope. It's actually one of the greatest and best motivations there is: the knowledge that there is always more to discover, more work to do, and (because we're not all saints here) more chances to put your name to something remarkable. Discovery will always be open-ended. Scientific research continues to thrive exactly *because* of what we can't prove, and probably never will. After all, how boring would life (and science) be if we'd figured out all the answers already?

8. Bias

How to be (more) objective

In many areas of life, we accept bias as inevitable. We are biased about the sports teams we support and the political parties we vote for, about the places where we grew up, our friends and our children. We are made so by some combination of the way we have been raised, the influences we are surrounded by, and the outcomes we wish to see. (And believe me, when you grow up autistic, people don't hide their bias towards children who behave in any way they deem less than 'normal'.) Often we find ourselves taking a side and sticking to it, even though part of us knows that not all the evidence is in our favour. We back up someone or something more because personal allegiance trumps everything else. It was never a penalty. He's not that kind of person. My daughter would never do that.

Science should be the perfect antidote to this: a pursuit in which we don't wear the shirt, shout at the referee or complain that no one else gets it. It is meant to be a place of reason over feeling, evidence over emotion, and answers that we listen to even when they aren't what we want to hear. Science is meant to be the stoic practice of objectivity. That's one of the reasons I always loved it: it felt like a way of imposing order on a world that didn't otherwise make sense to me. Its cold certainties were the anecdote to the heated chaos I often felt trying to navigate life.

But not so fast. Just because scientists have to gather evidence and show their workings doesn't mean they are somehow free from sin when it comes to bias. In scientific work there is ample scope to fall in love with your own theory and fit the evidence to it, to disregard a good idea because it doesn't fit with your priors, or to back the argument of someone more senior and established simply through deference to authority. All of these biases may be unconscious, to the point where researchers put their fingers on the scale without even realizing (and even if, unlike me, you are not cack-handed and a bad lab scientist).

Science is no more a bias-free zone than any other walk of life. Despite all the funky charts, detailed formulas and sometimes abstract language, it is as much a human activity as watching a game of football. Scientist is just another word for a person who questions the world and wants to discover how something works. Who pokes around until they find something enlightening, so they can make discoveries, get published, advance their career, win prizes and maybe even change the world a little. A scientist is also someone liable to be influenced by the individual or institution that is paying for their work, the journal to which they will submit their next paper, and the work being done by others in their field. Pretty much the same petri dish for breeding bias that we are all ultimately stuck in as human beings. (We even have our own version of swearing at the ref, which involves the contents of your blind peer review and a very private room.)

All the rigour that goes into scientific work can't free it from the clutches of subjectivity. However hard we try, science will never be completely free from human bias. But it does offer an excellent lens through which to examine bias: how it operates, consciously and unconsciously, the challenges it presents and perhaps even the help it can provide. If we accept that a degree of bias is inherent to how we think, work and make decisions, we are left with the question of what to do with it. Should we try and reduce that bias, challenge and counterbalance it, or simply account and allow for it? Is bias the enemy to be hunted down and minimized, or something we need to accept and even seek to take advantage of? Ironically it is science, the supposed paragon of objectivity, that provides one of the best ways to understand the murky and inconvenient reality of bias in everything we do.

Every scientific experiment ever conducted has been in some way biased, and for a simple reason. We – scientists – have been involved in it. The way an experiment is conceived, designed, measured and carried out, the way its results are interpreted and presented, are all places in which personal and technical bias can and does creep in. Even by setting up an experiment in the first place, a scientist is essentially making a statement about what they hope or expect will happen. The researcher is looking at or for something, which immediately disposes them against inputs they haven't considered or results they didn't anticipate. It's for this reason that, after much agonizing, I chose to do a PhD in bioinformatics, an interdisciplinary field that felt less limiting to the work I could do and the techniques I could use. I could see it all.

In simple terms, as humans we have a tendency to find what we wish to see. While scientific rigour should guard against this, even trained professionals can fall into this trap, with results ranging from mistaken observations to dubious interpretations and experiments that were skewed by their very design. These problems can be illustrated by three well-known scientific misadventures involving animals: a chicken that never existed, a horse that wasn't what it seemed, and some rats that weren't as different as they were made out to be. All are examples of the spectrum of bias, running from wishful thinking to motivated reasoning, that can mess with the process of scientific inquiry.

Let's start with the chicken (or the egg). In 1999 a fossil emerged in north-east China that appeared to answer a question palaeontologists had long been pondering: how birds as we now know them had developed. Where had they come from, how had they developed the ability to fly, and what did that have to do with dinosaurs? Biologists had long belived believed there must be an evolutionary link between dinosaurs and birds, but had so far lacked the evidence to demonstrate it. Then, almost miraculously, it arrived. In October that year, the National Geographic Society announced the discovery of a fossil that appeared to show the physiology of a flying bird, plus the unmistakable tail of a dinosaur. In a Press conference and subsequent *National Geographic* article, the find was heralded as a 'missing link between terrestrial dinosaurs and birds that could actually fly', exhibiting characteristics that were 'exactly what scientists would expect to find in dinosaurs experimenting with flight'.[1]

Note the phrase 'expect to find'. They had expected, they had found, and moved quickly to put the two together. *National Geographic* rushed to publish the discovery, but had to retract only months later. It had indeed been observing the body of a bird and the tail of a dinosaur, because the object they had been studying was a composite, two real fossils from completely different animals (and different species) stuck together. A Chinese palaeontologist, Xu Xing, even managed to locate the other half of the dinosaur fossil that had provided the tail.[2]

A much publicized advance in the history of avian life had fallen to earth with a bump. Ironically, the red-faced researchers had been on the right track: the link between dinosaurs and birdlife would subsequently be demonstrated with evidence that hadn't been stitched together in Dr Frankenstein's lab.[3] Yet in their pursuit of a hunch that would ultimately be proven right, they had fallen into one of science's biggest traps: allowing expectation to run ahead of evidence, a tendency compounded by the desire to be the first to print, which can lead to corners being cut or inconsistencies not being checked. A bit like deciding you are in love with the person you have just started dating, so you have someone to bring to your sister's wedding.

This desire to find what we are looking for can skew the way scientists interpret evidence – the observer-expectation bias. It can also create flaws in the research itself, through experimenter bias, where the person conducting the experiment inadvertently influences the elements involved in it. Consider a strange quirk of the 'Viking' mission to Mars conducted by NASA in 1976. This involved landing two craft on the Red Planet to conduct tests on the soil and atmosphere and establish whether they did or could support life forms. The consensus at the time, and since, was that they confirmed the view that nothing could survive in this punishing environment. Despite some anomalies in the findings, they were clear enough for one of the project's scientists to conclude: 'No bodies, no life.' Recently that has been questioned by the astrobiologist Dirk Schulze-Makuch. He has argued that the original researchers skewed the results by adding water to the soil as part of their experiments, using the same methods they would have done on earth, and in the process potentially destroying microbes that might have adapted to survive in this famously arid environment.[4] This is a system bias of assuming that all life forms depend on abundant water.

It is only a hypothesis, but even without any experimental grounding, it helps to illustrate experimenter bias – the way researchers may undermine their own work through the approach they take to it, and the inadvertent influence their methods may have on the subjects of their experiment. A famous example of this was Clever Hans, a horse that became famous in 1900s Berlin because it could apparently count, tell the time and spell out words (tapping its foot to indicate numbers or letters of the alphabet). But this was no equine prodigy, simply a horse that responded to the body language of its trainer, picking up its cues for when to start and stop tapping.[5]

This case of Clever Hans shows us that experimenter bias is especially prevalent when dealing with animal subjects, just as it is with humans. The psychologist Robert Rosenthal set out to demonstrate this in the 1960s, courtesy of a group of his students and a cohort of albino lab rats. In reality, these were completely normal rats like those regularly used for research purposes. But not as far as the students participating in the study were concerned. They were told that half of them would be working with 'bright' animals that had been bred to be skilful at escaping from mazes, while the others would be working with 'maze-dull' specimens. Having deliberately misled his experimenters in this way, Rosenthal sat back and waited to see if it would make any difference. The results were clear: the rats being handled by those who thought they were working with the clever cohort performed much better – almost twice as well – as those that had been labelled 'dull'.

Rosenthal had shown that influencing the person doing the experiment can have a measurable impact on the result, and that changes in the behaviour of the animal subjects 'can be systematically induced and demonstrated'. By telling the handlers that they were dealing with smarter animals, they became more positively disposed towards them, handled them more delicately and achieved better results. That process became self-fulfilling as the 'bright' rats continued to respond well to the positive reinforcement of their successful navigation of the mazes, which may have been communicated to the animals through unconscious messages such as warmer or drier hands.[6] The psychologist achieved similar results in an experiment with elementary school teachers, who were told that some of the children in their classes (in fact chosen at random) had been identified by a Harvard test as 'growth spurters' showing particular aptitude and potential. They duly did better – in terms of improving their IQ score – over the following year, especially the youngest children aged six to eight. It was, Rosenthal wrote, 'further evidence that one person's expectations of another's behavior may come to serve as a self-fulfilling prophecy'.[7] (Not a conclusion that anyone who was labelled 'naughty' or 'difficult' at school needed to have confirmed through a scientific study.)

In other words, bias does not just exist on the sidelines, as a form of prejudice without consequence. It also makes a measurable difference to behaviour, relationships and everything that follows from them. As any woman who works in science will tell you, such preconceptions are littered through our field. What people expect of you as a woman in science can often determine what they decide about you, regardless of the evidence. Studies conducted in the 2010s found both that men were twice as likely to be hired for a mathematical job when gender and appearance were the only known factors, and that identical scientific papers were more likely to be deemed high quality if given the name of a male author.[8]

While bias may be subtle and often operate unconsciously, its influence can be profound. The suggestibility of the human psyche means we can be influenced without realizing, our behaviour and thought processes altered subconsciously. These influences don't just exist within the confines of a scientific experiment. They also surround every scientist who has ever written a research proposal, liable to be steered as much by what they believe a funder or journal will expect, the work others in their field or lab are doing, and what will be good for their career, as by what is most scientifically interesting or important. This isn't some big conspiracy: grant funders aren't evil or trying to suppress good work, but the reality is that any scientist working within the system must contend with its biases about what is and isn't a worthwhile research programme.

Bias is all around us, therefore, no more or less prevalent in science than any other part of life. Each of us as an individual harbours bias, and so do the systems we work in. Any organization ultimately functions according to the accumulated cognitive biases of the people within it: they will generally gravitate towards other people who look or think how they do, and to doing things based on what did or didn't work before. An entrepreneur who has built one successful business is likely to think they can repeat the trick (and to get investment to do so), even if none of the circumstances that allowed them to do it the first time still apply. Just as a company that tried and failed with a radical reorganization or product launch will be less inclined to try something so bold again, while people are there who still remember how it all went wrong the last time.

It's the same in science, where researchers may become once bitten, twice shy if an experiment in a certain field or using a particular technique goes wrong – in some cases wrongly settling for less because they mistake bad luck for error. In my PhD research, where your daily duty is to question everything about your field and in part your general existence, I would often fall into the tempting trap of taking hits personally, because you care so much. It even got to the point where I didn't want to do anything because I didn't want to be biased, from the food I ate to the side of my bed I woke up on in the morning. To take hits is part of the job, but also not letting them cripple you. This is something I had to learn pretty fast.

Equally, researchers may assume that what worked for them once will continue. And it goes on, through our personal lives, hobbies and willingness to take risks and try new things. So much of human existence derives from these availability biases, as we default to the norms of what our lives have led us to believe is normal or desirable. When someone says: 'This is what I think,' or 'Experience leads me to believe,' or 'Everyone knows that,' they may as well be saying: 'These are my biases.' So often we dress up individual or institutional experience as universal truth and knowledge, rather than recognizing it for the limited and faulty data set that it is. In your career you will hear from all sorts of people who have made it to the top and shower tips and advice on those still making their way. Some of this may well be useful, but the real insight is probably mixed in with all sorts of thinking that is either out of date or non-transferable from the circumstances in which it happened to that individual. Remember survivor bias the next time someone pops up telling you how to plan your career, or the secret to a happy relationship.

Can (and should) we do anything about this prevalence of bias throughout our lives? Is bias something we should seek to address and counterbalance, in the hope of achieving something closer to neutrality? In some important ways, yes. The practice of 'double blind' research was developed, especially for use in clinical trials, to counteract the kind of subconscious signalling that was revealed by the story of not so Clever Hans. In these experiments, neither researcher nor participant is allowed to know, for example, whether they are receiving a potential new wonder drug or a placebo. Similar protocols are followed in forensic laboratories: when examining fingerprints in a criminal investigation, technicians are drip-fed information about the sample they have been asked to work on. Studies have shown that, if they have all the evidence and background context at once, it can introduce bias – for instance, becoming more likely to assess a match between two fingerprints, if they know it will help to make the case against someone who is believed to be guilty.[9] It is also good practice for people to document their findings and analysis at each stage, to create an accountable record of what the evidence shows, and avoid situations where a judgement is subconsciously altered in the light of new information that does not directly alter the initial finding (but may cast it in a new light). A good tip for us all: it's only human to edit the record of what we think about something with the benefit of hindsight, once the game has been played and the result set in stone. There can be a huge power to writing down what you think in real time – whether about goals for the year, a career change, or a new relationship. It may surprise you to be reminded of how differently you perceived something six months or two years earlier, but this kind of accountability is how we learn and avoid letting our subconscious trick us into believing we always felt this way.

Mitigating bias is something most scientists work carefully to achieve, whether through the sampling techniques used to select a representative group of participants for a trial, the use of control experiments, methods such as double-blinding, or measures taken to avoid an algorithm overfitting to its training data. (Yes, even our trusty algorithms are biased – if anything, they can perpetuate the biases already in the data collected!) This is why statistics and their use in algorithmic predictions should be met with caution. There is also greater awareness of the issue of bias outside science, in areas such as recruitment where faulty human judgement is to the fore. The psychologists Daniel Kahneman, Olivier Sibony and Cass Sunstein coined the term 'decision hygiene' to describe processes that seek to avoid either bias or what they call 'noise' from affecting such judgement calls – noise being the general existence of variation and inconsistency in decision making, whereas bias is skewing people in a particular direction. Examples of this include providing information in a set sequence, as described above, or requiring job interviewers to follow a more structured and inclusive process – grading candidates against an agreed list of requirements or asking a set series of questions, versus simply having an open conversation and providing an overall assessment that is more liable to bias.[10]

They liken this to handwashing. Because noise is caused by so many different factors – they show how decisions in a court of law can be affected by everything from whether the local sports team won or lost at the weekend to the temperature outside that day – you cannot simply correct for one factor. You have to try and wash your hands of all of them – 'adopting techniques that reduce noise without ever knowing which underlying errors you are helping to avoid'. (Whereas with bias, you know what you're looking for and can make more specific interventions.)

Can we wash our hands of bias and the factors that create so much variance in our decision making? To some extent, the answer seems to be yes. A meta-analysis (study of studies) in this area has found that job interviews done without any set format or requirements 'were significantly more susceptible to bias than were structured interviews'.[11] That is despite the fact that interviewers much prefer being able to set the terms of the conversation themselves, and decide what they want to take from it, i.e. following their biases. In fact, it turns out that the people doing the interviews have a belief in their own capabilities that far outstrips reality. (Who knew?) Kahneman cites an experiment in which interviews were conducted using only 'yes' or 'no' answers, interviewees were instructed to answer randomly, and the people interviewing them said they could 'infer a lot about this person given the amount of time we spent together'. This, he suggests, demonstrates how easily self-confidence can warp human judgement. 'As we can often find an imaginary pattern in random data or imagine a shape in the contours of a cloud, we are capable of finding logic in perfectly meaningless answers.'[12] (A helpful message for your friend who thinks their crush is trying to communicate subconsciously with them.)

So while we like to trust ourselves to make decisions (Kahneman also gives the example of how American judges dislike guidelines that have been proven to reduce bias in the sentences they hand down), it seems we actually do better when following more of a box-ticking approach, which may be boring but does at least have the benefit of being consistent and accountable. Clearly there is a place for decision hygiene in situations where people would otherwise be subjected to highly variable human judgement that could affect their career, their health and even their liberty.

But the principle has its limits too. While uniformity can iron out variance in situations where decisions need to be consistent and not at the mercy of people's preconceptions, it is something we should seek to avoid in other scenarios. As individuals, our biases are too much a part of us simply to expunge. As a job interviewer or research scientist, we might engage in a process that helps us to limit them, but in so many other areas of our lives we need these inbuilt preferences, almost an emotional sat-nav for getting around and tackling big problems. Whenever I want to hone in on my biases, I often follow the self-made acronym of BIAS (Behaviour In Alternative Scenarios) to remind myself of my default assumptions and responses, and how they may not translate well to a new situation. As I have learned, it isn't about being free of bias, but knowing what your starting point is through simulating various scenarios (perhaps different audiences or situations or communities when forming an opinion), so you can respond intelligently and with humanity to others.

Bias provides context and shapes our understanding of the world. It is a force that can inspire creativity and original thought in abundance. So many great scientists, entrepreneurs and inventors have succeeded because they were driven in a particular direction by the experiences of their life and the dispositions of their mind. They can see what others cannot precisely *because* they are biased, and those biases give them a perspective that no one else can share – a rotating beam from a lighthouse that is hidden to everyone but us. To try and negate these impulses in the name of neutrality is not just difficult; it is like trying to cut out your personality, or simply not to exist. Unfortunately, many people whose biases stray across the tramlines of social norms have come to exactly this conclusion. They minimize and 'mask' their true selves to negate the criticisms of others. Like all autistic people I have done and still do this; at one point, I even thought that it would be better to live without emotion at all, trying to suppress everything I felt and wanted in the name of neutrality and social acceptance. Unsurprisingly (in retrospect) it was one of the most miserable experiences of my life.

So we should be careful with bias. There are occasions when people owe a professional obligation to be as neutral and consistent as possible, and here Kahneman's handwashing is an important way of ensuring this. But we should never try and destroy the bias that makes us human, with all the wonderful creative potential that contains. We should not try and deny our personalities or sacrifice our unique perspective for the sake of appearing professional. Inclusion is not about people fitting themselves to the norm, but the norm expanding to make room for them. So while employers have an obligation to treat everyone the same in some important regards, they should never expect everyone to be or think the same way. There is a time to try and drive out the bias and quiet the noise that surrounds human judgement, but also a time to let those things loose, and allow our biases to lead us wherever we go next.

The question of what to do with bias is also more complicated than whether we need to swat it like a fly or use it as creative fuel for our work. So deeply embedded are biases in human psychology that they also provide valuable insight into how we think and behave, as individuals, teams and organizations. Bias is something we cannot live without, even if we wanted to. It's also a lens through which we can understand not just ourselves but the people around us, working in the same teams or on similar problems. So what if, rather than simply treating them as the enemy or something innate, we also started to consider how we might put them to good use?

★

If scientists have a collective bias, it is that they like to understand how things work. We were the kids who asked too many questions and liked to build things and deconstruct things, because the world was just too interesting to be left alone. At its best, a career in science allows us to carry on doing exactly that, poking around and exploring cool stuff (with added grant applications, departmental meetings and job interviews, it must be admitted). But what if that most scientific of instincts wasn't always as helpful as it seemed? What if the urge to explore, discover and make sense actually got in the way of finding things out? And what if worrying less about what we really know or understand could be the shortcut to scientific nirvana?

This is (in part) the argument of Dr Alexander Wissner-Gross, a Harvard Fellow who is one of many physicists working at the juncture between human and artificial intelligence, and a researcher who has taken an unusual approach. Rather than beginning his quest to understand intelligence with the human brain – trying to decode its mysteries and then find a way to transpose the learnings to an AI model – he has started with the machine. His focus is simple: to create a working model and then use it to explore fundamental principles, rather than first elucidating the principles and using them to build. It's about 'thinking from a phenomenological perspective as opposed to a mechanistic perspective', he suggests. 'The phenomenological perspective is: "What's the effect on the world?" And the mechanistic effect is: "How do all the little pieces inside of the system work?"'

'There's a rich history of this in science,' he explains. 'If you look back at the history of thermodynamics, arguably we needed artificial thermodynamic cycles in the form of steam engines before we had a mature scientific understanding of thermodynamics. It wouldn't surprise me at all if a mature understanding of neurobiology ends up being largely informed by first achieving artificial general intelligence.' In other words, don't try and solve the scientific problem that helps you to build the machine, but build the machine and watch the scientific problem solve itself. As he puts it: 'Generally I know what I'm trying to build, and science falls out almost for free on the way to engineering the desired outcome.' Which in the case of artificial general intelligence, means: 'The right question is not "What is intelligence?", but "What does intelligence do?" '[13]

By taking the phenomenological approach, he arrived at an answer: 'Intelligence doesn't like to get trapped. Intelligence tries to maximize future freedom of action and keep options open.' This was demonstrated by a software program that was applied to tasks subjected to 'casual entropic forcing' – an environment where disorder is maximized, and those options are kept as open as possible (ironically akin to the approach used in cancer to adapt and grow through producing genetic noise). The results were impressive. Presented with the scenario of a cart on two wheels and a metal pole – a standard test in training humanoid robots to walk – the algorithm successfully balanced the pole upright, even though it had been given no instructions to do so and no objective to achieve. In similarly open-ended situations, it was able to direct theoretical container ships through the Panama Canal (without knowing that it existed) and to make money trading financial markets.[14]

The fascinating conclusion seemed clear: through exposure to disorder and chaos there is a clear correlation between the ability of a system to be adaptive and to consider all possible options, and to take actions that we would typically associate with intelligence. The more open-endedly something 'thinks', the more intelligently it is likely to act. This idea had been arrived at not through the bottom-up approach often associated with scientific research, of establishing known parameters and building out from them, but of building a model and playing with it. It is this phenomenological approach, Wissner-Gross suggests, that is likely to bring about the advent of artificial general intelligence – algorithms that can think just as well as humans. 'If your goal is to accelerate superintelligence, then maybe it's a more efficient approach to just focus on creating systems that – by whatever mechanism – have the same effect as intelligence or superintelligence on their environment, without worrying too much about mechanistic theories or microscopic theories of how existing biological intelligences work.'[15] We might equally conclude that sometimes in life we just have to get on and do things, even if we feel underprepared. You can spend for ever trying to weigh the pros and cons and to come up with a perfect formula for what equals happiness in your life, or you could just do the things that instinct draws you towards, accept there will be good days and bad days, and let the formula worry about itself.

I love the idea of causal entropic forces not just because they illuminate an area of science close to my heart. They are also an example of what can be achieved by going against the grain, understanding the biases at play (in this case, across an entire field of research) and using those almost as signposts to explore a new approach or re-examine a consensus that has gone unchallenged for too long. It is almost certainly true that the mechanistic approach, as Wissner-Gross defines it, is the prevailing bias within physics. Physicists, and in particular engineers, love to take things apart and see how all the cogs fit together; they are natural disassemblers and tinkerers, never happier than when being presented with a system and asked to figure out how it works. Therefore the phenomenological approach is by definition likely to lead researchers into places where fewer people are thinking and working. It is hacking the system, using institutional biases as a negative image to reveal ideas and approaches that are not being pursued by the mainstream. Some entrepreneurs will talk about trying to 'zig when others zag' – a hipster doing something unusual and unexpected that may get people interested, rather than just imitating what already exists. This can be good advice in science too, but to follow it you first need to understand what all the zagging is about – what most people are doing and why, the assumptions and thought processes that led them to this point. You need to study the herd before you can distance yourself from it. Or put another way, you have to understand its biases. Which is where being different gives you a head start, since you are used to being apart from the mainstream, studying it from a distance and knowing its strengths as well as its weaknesses.

It is the understanding of bias that holds the key both to avoiding its pitfalls and harnessing its power. We are all individuals with almost as many cognitive biases as we've had hot dinners, and there is no prospect of us behaving in an entirely unbiased manner, nor living in a world that is ever close to neutral. The only question is how to make the best of all this impulse, preconditioning and assumption, at the same time as avoiding the worst of what it can do. Science shows us that where we can identify and isolate a specific bias, it can be controlled through counter-measures, while if we know a situation is liable to be skewed by noise, then tools of process and standardization can help to even the score.

Understanding bias can also help us to make use of it. By understanding our own biases, other people's and those of the system we are working in, we are ironically better equipped to behave in a way we might call 'unbiased' – creating genuine inclusion through supporting people in an equitable way, according to their particular needs and circumstances, rather than with the broad brush of equality, in which we assume everyone must be treated equally. While standardization has its role in Kahneman's noise hygiene, the creed of equality has its limitations, one which an appreciation of biases (both individual and systemic) can help us to overcome.

There is also a personal benefit to learning about and exploring our biases. By doing so we learn about ourselves – the influences that have shaped how we think, the advantages this may give us and the limitations it may impose. Reframing the statement 'This is what I think/know' to 'This is my bias' turns a static statement of belief into a fluid one that recognizes one disposition while leaving us more open to alternatives. It facilitates flexible thinking and encourages empathy with other points of view. It furthers the lifelong process that is understanding who we are as a person, the forces that have shaped us, and the factors that explain why we think, behave and react as we do.

Saying that we are biased is not some embarrassing admission but a simple statement of fact about what it means to be human. Any scientist knows that an understanding of their own biases, the biases of their field of study and of the institution they work in, is crucial to their work. The same should be true in every area of life. Rather than demonizing bias as the enemy of objectivity, we should be embracing it as the key to understanding how and why people think and behave. If we want to understand ourselves, our friends, the people we love and those we work with every weekday, then we need to get in tune with bias. Doing so makes us more self-aware, more empathetic, and more liberated to be ourselves. The rich tapestry of human bias is the fabric of life itself. So pick up your needles, and don't be afraid to add some threads of your own.

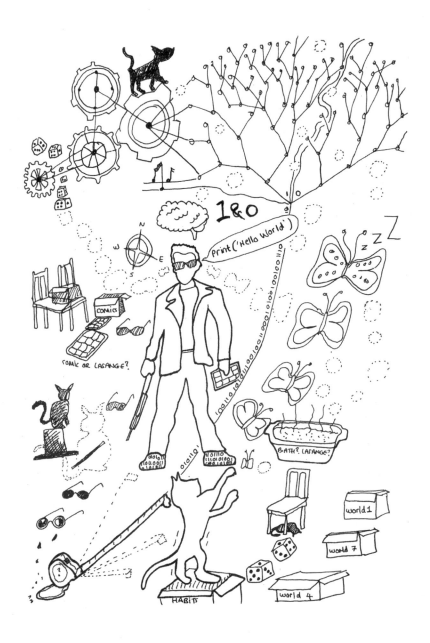

9. Imagination

How to build worlds and expand reality

As artificial intelligence becomes more prevalent in all our lives, the inevitable question follows: Where does this end? Will it take us to a crossover point at which 'intelligent' machines become more powerful than their human programmers, rendering us all the servants of technology? Will we become victims of algorithms that supersede us and robots that could eliminate us, Terminator style?

This is far from a new fear. At the time of writing, Arnie's first outing as everyone's favourite, and extremely toned, cyborg is almost forty years old, and it is more than two centuries since the original Luddites were destroying equipment in textile factories to protest against technology taking their jobs away. The feeling that humans might be sowing the seeds of their own destruction through the machines they build stretches back to some of our oldest literature: to the Greek myth of Daedalus and Icarus, flying too close to the sun on artificial wings, and the biblical allegory of the Tower of Babel, another tale of soaring human ambition that overreached itself. The entirety of human existence has been based on trying to catch up with our imaginations, but in the process we have already manufactured things powerful enough to destroy our species, and with advanced AI we may be adding to that list.

But what if this long-running debate was actually moot? If the battle between human and machine had already been fought and the story told? If, rather than being the guinea pigs for how humanity will fare in a world with increasingly powerful computer systems, we are in fact the creations of an epoch in which that question was already answered, a long time ago.

That is the notion put forward by the philosopher Nick Bostrom, who in 2003 proposed what has become known as simulation theory: that we may all be living in a version of reality created by our descendants, whose access to unthinkably powerful computing allows them to create simulations involving people who act with what feels like conscious thought.[1] This would make us not the manifestations of biological cells but lines of code, ones and zeros in the place of atoms and proteins. It would mean we are simply characters in a computer game far more complex and intricate than we can imagine creating ourselves (while also making an excellent riposte to people who say things like 'You couldn't make it up!'). For these hypothetical future beings, we may be no more nor less meaningful than avatars in *FarmVille*.

Like technophobia, the idea behind simulation theory is far from a novelty in human history (or, if you prefer, this particular playthrough of the game we're in). Since philosophers have been recording their thoughts, they have been pondering the nature of reality versus illusion. The ancient Greek philosopher Plato, in his allegory of the cave, described prisoners fated to spend their lives facing a wall, watching only the reflected shadows of objects passing behind them. His seventeenth-century counterpart René Descartes envisaged a 'malicious demon of the utmost power and cunning [who] has employed all his energies in order to deceive me. I shall think that the sky, the air, the earth, colours, shapes, sounds and all external things are merely the delusions of dreams which he has devised to ensnare my judgement.'[2] Around the same time as Plato, *c.*300 BC, the Chinese philosopher Zhuangzi pondered the experience of having fallen asleep and dreamed he was a butterfly. 'Suddenly he woke up and there he was, solid and unmistakably

Zhuangzi. But he didn't know if he was the Zhuangzi who had dreamt he was a butterfly, or a butterfly dreaming he was Zhuangzi.'[3] A bit like when you can't remember whether you're losing sleep because you're stressed about something, or stressed because you aren't sleeping enough.

As these constructs suggest, we don't have to poke too hard at the foundations of our existence and consciousness before things begin to feel a bit wobbly. Start to look around and suddenly the world doesn't seem to be quite as firm as the ground beneath our feet. There are optical illusions, paradoxes of nature and gaps in our understanding which all erode our sense of certainty – the belief that the world around us is something we can know and understand with precision, like the recipe for baking a sponge cake.

For scientists, such recipes are the basis of everything. Yes, the theory we are working with may be incomplete, unproven or subject to being improved or overturned in the future, but it is at least a stepping stone on which to stand – one of a series that can help us to inch towards new ideas and discoveries. Scientists are comfortable with uncertainty, but they also thrive on the tangible: a world and entities in it that can somehow be observed, broken down into their constituent parts, studied, modelled and presented like the layers of a lasagne. Most of the science which this book has discussed ultimately arises from that frame of thinking: that the unknown can be made at least a little more knowable through observation, experimentation and measurement. In simple terms, we like things we can either confirm or disprove.

But there is a branch of science for which this is insufficient, and simply unworkable. At the frontier between physics and philosophy, people tangle with ideas that are mind altering and brain bending: from Bostrom's simulation theory to the idea that our world may not be singular, but one of infinite parallel realities, and the thought that our concept of time passing is an illusion, with no such thing as the present linked directly to past and future. At this point you might tap the table in front of you and say it seems solid enough to you, and why worry about anything else? But it is exactly these ventures into what sounds like science fiction that help knowledge to progress – forging fantastical views and conceptions of the universe that may sound far out, but are also an incentive for research to close the gap between what we can imagine and what we actually perceive. It isn't an either/or matter, but one view constantly catching up to meet the other. To enter this weird and wobbly space, we first have to throw off the comfort blanket of what is tangible. We have to give up some of the tools that science has taught us are essential.

This is where theory becomes twisty and scientific discovery speculative: those stepping stones get more slippery, and move further apart. It's a branch of scientific inquiry that forces us to question the nature of our reality, the underpinnings of our human consciousness, and our assumption that we live in a world that is at its core observable. Whereas so much of science is concerned with building a picture through experimentation and evidence, this branch of theoretical physics is as much about tearing down our reality – pulling at loose threads in our theories of the universe and watching whole bits of the picture unravel. It's where scientists start to wonder if our conceptions of what's out there even come close to reflecting the true nature of reality. And it's the point at which science dares to ask: What if everything you thought you knew about the world was probably wrong?

Before we can take our first steps into that brave new world, we first have to turn back, around a century in time (if time exists) to the development of quantum mechanics. Those two words are enough to send a shiver down the spine of people who do this for a living, let alone those only vaguely familiar with the concept. As the physicist Sean Carroll, who we met in Chapter 4, has written: 'Quantum mechanics has something of a reputation for being difficult, mysterious, just this side of magic.' Yet it's also, he suggests, 'the deepest, most comprehensive view of reality we have'.[4]

The mysterious reputation of quantum mechanics is itself something of a paradox, for the concept is both rock solid and a foundation of our entire modern way of living. Without quantum theory's ability to explain the behaviour of particles at the subatomic level, we would not have been able to model the movement of electrons that allowed us to build the semiconductors that power the computer I am writing this on, the mobile in your pocket, or pretty much any electronic device on which we rely. It is no exaggeration to say that life as we know it is dependent on quantum mechanics. But nor is it over-egging the pudding to suggest that this field of science strikes more fear into practitioners and onlookers than any other.

Let's take a deep breath and dive in. At its core, quantum mechanics overturns the belief of classical physics that every particle is a singular observable entity, whose position and movement can be determined according to Newton's laws of motion. Quantum mechanics delves a level deeper, into the behaviour of the bits within the atom – beyond reality as we see it. It shows that, at the subatomic level, a particle (such as an electron or photon) is not confined to a singular position. Instead of being plotted as a point on a map, it is expressed through what we call a wave function – essentially a spread of the locations and energetic states in which it might be, a statement of probabilities rather than any absolute certainty.

This may seem like a piece of scientific nit-picking – a Spot the Ball competition for people too fond of equations – but its implications have been profound. Crossing the threshold of quantum mechanics has allowed physicists to reach a number of conclusions so radical that they have altered our entire understanding of the world. One was that a particle is not a fixed quantity, tied to a particular location in a given moment. It also has what is called superposition – the ability to exist in more than one place at the same time (the basis for the thought experiment of Schrödinger's Cat, capable of being simultaneously alive and dead). Another was the potential for quantum 'tunnelling' – in which particles, thanks to their wave-like behaviour, can pass through apparently insuperable barriers (a concept fundamental to developing the diodes – switches – that govern the flow of current in a semiconductor). As well as tunnelling through what should be brick walls, under quantum theory particles can become entangled, linking together and sharing properties even though they may be positioned light years away from each other.

A further complication of quantum theory comes when the external observer – us – starts to interfere with all this happy confusion among subatomic particles. They are busy manifesting, tunnelling, entangling, and then a scientist comes along, brandishing a microscope or shining a light and trying to work out what's going on. The very act of this measurement has a remarkable effect on the particles under the microscope (real or metaphorical). It results in what is known as wave function collapse: the superposition capturing the probabilistic range of where the particle *might* be is immediately condensed to a definite state of its position at the point of it being measured. A bit like a game of micromolecular grandfather's footsteps – the moment the observer turns around and tries to observe the particles, they freeze (or at least appear to).

Remember that the wave function is an expression of probability: all the places our particle might be. Measure it and we reduce those possibilities into the appearance of certainty. That might sound like we are negating the entire purpose of the wave as an expression of a quantum particle's elusiveness, but not so fast. For the measurement – the point we have plotted on the map – is not so much reality as our only way of perceiving it. We cannot 'see' the superposition – multiple realities all at once, Schrödinger's cat both alive and dead – but that doesn't mean it did not exist. This conflict between the probabilistic state in which the particle exists, and the deterministic state in which we as humans observe, is known in quantum theory as the measurement problem. We can't study a subatomic particle without imposing the limitations of our human perspective onto it.

This gets to the point of why quantum theory is both so important and so difficult to grasp in conceptual terms. It is an operating system for everything that exists in the world that's completely robust in principle, yet completely at odds with how we intuitively think, observe and measure those things in practice. As Sean Carroll has written: 'The enigma at the heart of quantum reality can be summed up in a simple motto: What we *see* when we look at the world seems to be fundamentally different from what actually *is*.'[5]

This is the conceptual challenge of quantum mechanics: whereas scientists (and frankly most people) enjoy the certainty of things they can observe, measure and quantify, the quantum world is one whose building blocks are frustratingly fluid – in several places at once, defying apparent obstacles and combining in mysterious ways. And the harder we fight to impose a sense of linear order on it, the more quantum behaviour resists. Meet the Uncertainty principle, which states that we cannot simultaneously measure a particle's position and speed. Indeed, the more we manage to quantify one of these things, the less we can know about the other. As you try and pin one jelly to the wall, another slides even further out of reach. It's as if the universe wishes to taunt those who would decode its mysteries, negating the scientific methods and tools they have brought to the party.

Hopefully by now you're getting an inkling of why quantum mechanics is such a tough nut in the scientific world, one that is treated with something approaching reverence, and which is almost always explained through the lens of how difficult it is to explain. Quantum theory has also been the subject of heated debate among some of the finest scientific minds that ever lived, especially the leading lights of nuclear physics in the 1920s and '30s who helped to develop the ideas: figures we met in the story of nuclear fusion, including Niels Bohr, Albert Einstein and now Werner Heisenberg. For Bohr and Heisenberg – who developed the Uncertainty principle – the probabilistic and intangible nature of quantum theory was the point, demonstrating that particles behaved in mysterious ways beyond the scope of classical physics. Einstein took the opposite view: all this haziness, he thought, pointed to quantum theory being an incomplete and unsatisfactory (if not incorrect) explanation for particle behaviour. The recourse to probabilities did not satisfy him: 'God does not play dice,' he wrote in a letter to another scientist.[6] After debating the question with Bohr at a landmark conference in 1927, he embarked on a long and fruitless search for 'hidden variables' that could support a more precise approach of quantum theory in line with the classical conception that what we see is what we get.

If even Einstein struggled to some extent with quantum theory, we can forgive ourselves for feeling a bit overwhelmed by the ideas it contains. Yet it is more than worth the effort to accept such brainache, a little like the misery of a long-haul flight to a brilliant holiday destination. So much of science at the frontiers of possibility, where we imagine things far outside our observable existence and tease out radical possibilities, stems from the reality-altering possibilities of quantum theory. Learn to look through quantum's hall of mirrors and you will never quite see the world the same way again. It stokes the imaginative potential of science like nothing else, precisely because it forces us to question the most fundamental tenets of how we observe and think about the world. Take the physicist Carlo Rovelli's concept of relational quantum mechanics, which suggests that we shouldn't think in terms of objects at all, but only the relationships between them. '[At] the core of the physical reality, it's not particles, it's relational connections,' he has said. 'Each object is defined by the way it interacts with something else. So when it's not interacting, it's just not existing.'[7] A chair only becomes a chair, he suggests, because we interact with it as something to sit on.

And that's not even the good part. While the implications of quantum mechanics are dramatic enough at the microscopic level, they get even weirder and more interesting when we extrapolate what we've learned about electrons, wave functions and superposition to, well, the entire universe. That was the contention of the mathematician Hugh Everett in the 1950s. He argued that we couldn't just use quantum mechanics as a theory to satisfy the behaviour of particles we cannot see; we must also consider what it would mean for everything that is visible and tangible around us: the entire universe.

Arising from this, he proposed what became known as the many-worlds interpretation (MWI). Three decades after Bohr and Einstein had first tackled the measurement problem of quantum theory, Everett suggested a new way of thinking about it. He asked what would happen if we abandoned the notion of wave-function collapse, and stopped trying to condense the superposition of a quantum particle into something recognizable to the lens of classical physics (the approach of what was known as the Copenhagen school of quantum mechanics). What if, rather than allowing alternative realities to 'disappear' through wave collapse, we assume that they continue in parallel: alternate worlds and realities in which observers just like ourselves are watching a different version of the superposition play out, one that we can't see?

There would be no single version of events, but an infinite number of branching versions, all occurring simultaneously. In this concept, we, the world and everything in it are part of what Everett defined as a 'universal wave function' – the entire spectrum of possibilities for every particle in the universe, ourselves included. (Good news the next time you have to sit in a boring meeting: in many other universes, more than your mind is elsewhere.) We occupy and observe our own bit of the superposition, but by doing so do not negate all the other possibilities – the wave function does not collapse but simply laps onto the shore of all these different worlds. The reality we see is not singular, but simply the only one of a countless number that are happening, out of sight and mind.

It almost goes without saying that this was (and is) a dramatic interpretation with huge consequences for our understanding of the world. It was a big enough deal that a PhD student, as Everett was when he developed MWI, was challenging some of the biggest beasts in his field (a bit like taking on the final boss right at the beginning of the game). The concept he put forward was even bolder. It said that the world scientists had been struggling to understand and explain for millennia was barely a rounding error on the universal balance sheet. Humanity as it understands itself may not occupy merely one small rock in a sprawling cosmos, but one that is replicated across an infinite number of alternative universes. We may, in other words, be even less of a big deal than we thought.[8]

Yet as far as Everett was concerned, his theory was not so radical. It simply took quantum mechanics at its word – that subatomic particles do indeed exist in a superposition of different states. Rather than trying to reconcile this with classical methods of observation and measurement, his interpretation encouraged that this ambiguity be accepted. We should, it argued, accept the multiple possibilities inherent in quantum theory rather than trying to fit them into neat boxes. They were all happening, in alternate worlds that we couldn't experience or interact with. Each time quantum particles collide, another branch is created in this endless succession of parallel realities.

Everett's theory may be complex in its implications, but its conception was simple enough: he was not creating or altering anything, but simply acknowledging what quantum mechanics said should exist beyond what we can see, and indeed measure. By contrast, it was those who argued for quantum wave collapse who were putting their fingers on the scale, getting rid of all those inconvenient alternative possibilities. They were essentially imposing the world view of classical physics onto quantum systems too complex to be contained within it. Their rationalization was in fact a simplification.

The Everett theory was ambitious, and he was dogged in defending it. When another scientist questioned the idea of multiple worlds on the basis that 'I simply do *not* branch,' he retorted that the same critique had been made of Copernicus when he first posited that the earth orbited the sun. 'I can't resist asking: Do you feel the motion of the Earth?'[9] Ironically his correspondent was the theoretical physicist Bryce DeWitt, who would become a convert to Everett's way of thinking and helped popularize his work in the 1970s, coining the 'many-worlds' label, a catchier shorthand for the idea of a universal wave function.[10]

Yet his initial scepticism was the more typical response. In the early days of the theory, advocates were thin on the ground. Under pressure from his supervisor, Everett published an abridged version of the theory in his 1957 PhD – omitting most descriptions of what he described as 'splitting worlds' – and left academic work behind. In 1959 he met Bohr, the godfather of the Copenhagen school, but described the encounter as 'doomed from the beginning'. Neither was able to convince the other of the value of their approach.[11]

For many of those trained in the Copenhagen school, Everett's conception of multiple realities was too intuitively challenging to be taken seriously. Richard Feynman summed up the feeling that this was just too big a leap to be comfortable, even for theoretical physicists. 'The concept of a "universal wave function" has serious difficulties,' he told a conference in 1957, while Everett was still completing his doctorate. '[The] function must contain amplitudes for all possible worlds depending upon all quantum mechanical possibilities in the past and thus one is forced to believe in the equal reality of an infinity of possible worlds.'[12] In other words: you really expect us to believe *that*? Others fleshed out the critique, complaining that since the parallel realities could never be observed, Everett had presented an unfalsifiable and therefore unscientific hypothesis, or that he had erred by trying to impose reality onto the wave function, which had only been intended as a mathematical device.

The idea of many worlds is still debated today, with plenty of advocates and detractors. For some of us, it just feels like everyday life. If you are autistic, or even just a bit of an overthinker, then you are probably used to gaming out multiple realities of a commute or supermarket trip before you leave the house. Done in moderation, this has some value. If nothing else, it is a useful lens for thinking about our lives: every decision taken is pushing us down one road at the expense of several others. If we hadn't taken this job, or met our significant other at that moment, our life wouldn't have come to a halt – it would just have taken a different course, one we may or may not be able to easily imagine. We can run the same mental exercise before we take big decisions too: all the different worlds that may be created if we say yes to something, and all those that may occur if we don't. A helpful reminder that we don't live in a binary reality of right and wrong, but a world of endless possibilities opening up and closing off as we make choices.

And whether you like the many-worlds interpretation or not, its influence is hard to deny. It has helped inspire the evolution of quantum cosmology – the attempt to develop a quantum-mechanical picture of how the universe operates – and ideas such as the 'multiverse' theory of different universes that developed in their own 'bubbles' of spacetime, moving at different speeds from the common route of the Big Bang. It has also contributed to the theoretical foundations of quantum computing – systems of potentially vast power and speed that run on 'qubits', data increments capable of existing in a superposition whereas the classical computing 'bit' must switch between a 0 and 1. This allows them to do in parallel and at huge scale tasks that traditional computers could only manage sequentially. The field was in part inspired by the work of the physicist David Deutsch, a disciple of Everett, who in the 1980s put forward the idea of a computer that could test and prove the existence of parallel realities. He has continued to argue for quantum computing as 'the first technology that allows useful tasks to be performed in collaboration between parallel universes'.[13]

Plenty disagree, but what cannot be denied about the many-worlds interpretation is that it has provoked debate and fed much fascinating work. Its outlandish notion means that some will always dismiss it, but among advocates it has been mind-expanding, allowing them to consider new and equally challenging possibilities. Perhaps there is a lesson to draw on here. For most of us, the idea that we might be living across infinite realities is no more than science fiction. But the thought process that underpins MWI is still a helpful one – one that says, this consensus view doesn't work at all, and what would happen if we thought about things completely differently? It reminds us that sometimes we need to step outside the box of our habits, assumptions and beliefs to make progress. We might be feeling stuck in a rut at work, on a particular project, or in a relationship.

Carrying on as you are is probably going to perpetuate that feeling, or make things even worse. But doing something completely different – turning a situation on its head or trying something you would never normally do – can be like jump-starting a car whose battery has run down. Even if you never go back to the pottery class for a second go or play another game of ultimate frisbee, it will engage body and mind in new ways, and you will feel different than you did before. And that is not just a perception or placebo effect: the brain can actually open up new pathways in response to unfamiliar situations, writing new code to assimilate to fresh stimuli. That is the power of pushing ourselves into the new and unfamiliar – a licence that theoretical physics gives by default, and which we should all try and give ourselves from time to time. It doesn't have to be a world-changing idea or experience, just one that's big enough to test your mind, and expand our reality even a little bit.

The willingness to explore big, scary and potentially crazy ideas is essential for scientists who are trying to probe the most mysterious and least tangible frontiers of our universe. The ability to cast off convention may appear unscientific, but it can also be necessary to take the big or unexpected steps that traditional thinking does not allow. Science at its most creative has to ask the big questions and consider possibilities that are heretical to the consensus view or frankly scary to think too hard about. That is why imagination is an important and underrated tool for the scientist who explores reality. It provides the courage that allows us to see what isn't, envisage what could be, to build and open up whole new worlds that would never otherwise have been considered.

The imaginative power of quantum mechanics does not just equip scientists to look outwards, to the furthest reaches of the cosmos. It has also inspired a search that turns inwards, whose focus is microscopic but whose meaning is potentially vast: the quest to understand the human mind, the source of our consciousness, and even the make-up of what we might call the soul.

This is not, of course, a new question. It's safe to assume that the desire to understand what makes us human is as old as science itself. Yet as scientific understanding of our biology has become increasingly sophisticated, the desire has grown to understand how all of the mechanics of muscles and nerves, synapses and transmitters, can amount to something more than physical existence and basic cognitive function – to the experiences of thought, emotion, personality and self that we call consciousness. As the philosopher David Chalmers wrote in his 1995 book *The Conscious Mind*: 'We have good reason to believe that consciousness arises from physical systems such as brains, but we have little idea how it arises or why it exists at all. How could a physical system such as a brain also be an *experiencer*?' He suggested that neuroscience had gone a long way to solving 'easy' problems like how the brain responds to stimuli and processes information, while dodging the 'hard' problem: 'Why is all this processing accompanied by an experienced inner life?'[14] How, in other words, do we amount to anything more than a bunch of functioning biological machinery? What is the bit that makes us human?

So fiddly a question has this been that the combination of neuroscientists, psychologists, physicists and philosophers who have tackled it has rarely been able to agree on a definition of consciousness, or even of the value of exploring it. 'Consciousness is a fascinating but elusive phenomenon. It is impossible to specify what it is, what it does, or why it evolved. Nothing worth reading has been written on it,' wrote one psychologist in 1989.[15] Fortunately, others have been more open-minded about the question of what is really going on inside our heads. Ideas have abounded since Chalmers defined the 'hard problem'.

Several of these focus on the way information is shared between and integrated among different parts of the brain – suggesting that this can lead to complex and unified experiences that are more than the sum of their parts (consciousness being proportional, one school of thought suggests, to the degree of information integration in a system). Another – the theory of 'predictive processing' – suggests that our experience of consciousness is arrived at as the brain reconciles the inputs it receives with the predictions it has made about something. The brain effectively guesses what something will look, taste or sound like, then receives sensory data via our various input mechanisms, and ultimately narrows the gap between the two. 'Perception, in this view, isn't a passive registration of external reality. It is an active construction, a kind of "controlled hallucination", in which the brain's best guesses are tied to the world – and the body – through a continuous process of prediction error minimization,' the neuroscientist Anil Seth has written.[16] (Controlled hallucinations would certainly explain why my celery soup always tastes like old socks whichever recipe I follow.)

Others have sought to apply quantum theory to the problem. If this is our more fundamental way to see the universe, perhaps it can help reveal the mystery of what is going on upstairs as well as out there. The primary advocate of this is the mathematical physicist Roger Penrose, winner of the Nobel Prize for his work on the 'singularity' in black holes, the point where density becomes infinite and the theory of general relativity ceases to apply. Even by those standards, his forays into the field of consciousness have been far out, pushing at the limits of what we do and can understand. His contention is that microtubules, proteins within the neuron, are capable of operating in quantum-mechanical ways, storing and processing information. Under the catchily named theory of Orchestrated objective reduction (Orch-OR), it is a process akin to wave-function collapse among the microtubules that creates 'a moment of conscious experience'. Much as the process of measuring a quantum particle reduces its superposition to something concrete, this theory suggests that our conscious experiences are the momentary distillation and simplification of information that is being crunched at unimaginable speed, deep in the substructures of our brain. The real quantum computers are in our own heads.[17]

This is a theory with more sceptics than supporters, with the most obvious questions surrounding how quantum effects could be experienced at body temperature, when they are typically associated with extremely cold conditions. But, as Penrose himself has argued, the value of the theory is more as a signpost than a destination. It is less a complete explanation of consciousness than an indication of where we need to go in order to improve our understanding: 'Whatever consciousness is, it must be beyond computable physics,' he has said. 'My claim is much worse, much more serious, much more outrageous than "It's quantum mechanics in the brain." It's not that consciousness depends on quantum mechanics, it's that it depends on where our current theories of quantum mechanics go wrong. It's to do with a theory that we don't know yet.'[18] In other words, we aren't smart enough to be able to decode the mysteries of the brain and human functioning as things stand. Our ability to map, model and measure every functioning piece within the brain is not going to lead us to complete answers about the human mind. Doing so will instead involve developing knowledge and conceiving of possibilities that are currently well beyond the scientific consensus and traditional physics. We will have to think bigger, and broader, than we are currently capable of doing and reformulate the nature of scientific theory itself to capture what we can't yet rationalize.

Whatever the ultimate value of the Orch-OR theory, it is a remarkable feat of the scientific imagination. It illustrates the lengths to which scientists must sometimes go to explore ideas so slippery and evasive that they may defy definition. Science often involves developing theories to explain how things work or might work. But as the ideas in this chapter – from simulation theory to the many-worlds interpretation and the quantum theory of consciousness – have shown, sometimes the requirement is to build entire worlds of possibility, just to start poking around in. These worlds might prove to be dead ends, or passageways to something better, but that does not make them useless. They are in fact an expression of one of the most important scientific skills: the ability to arrange the available pieces of a puzzle into a form that indicates those we may still be missing. Not just to see what is there but to perceive what *might* be; to understand when what appears to be a solution is actually a simplification that misses the real beauty of how something works.

Ideas like creativity and imagination are not usually associated with scientific research, and even some scientists may prefer not to embrace them. The thought that science can be a theatre for world-building, only one step removed from the storytelling of science fiction, runs counter to the idea that science is about what we can observe, measure, quantify and explain: a world of tangibles and firm foundations. By contrast, it is an exciting prospect for people who don't see a natural place for themselves in the world as it is, and often find an outlet through manifesting what a different reality would look like by creating science fiction or comics, in which a character just like them plays a central role and their voice is truly heard.

But the imagination isn't just about fictionalizing what a different, perhaps better, world would look like. It's also a fundamental tenet of pioneering research. As the principles of quantum mechanics show, sometimes the basic techniques of science get in the way of understanding. Their desire for rational order and distilled theory is insufficient to grasp the magnitude of what is going on in our universe. They obscure a complex reality by imposing a structure that is too restrictive to contain its full beauty and complexity. At that point, scientists (and frankly all of us) must learn to get comfortable releasing the straitjacket. Rather than trying to fit a framework around a problem, we must build a world big enough for it to live in, and get comfortable exploring all the glorious uncertainties and possibilities we have now imagined.

Conclusion

Why less may be more

Having travelled to the furthest frontiers of theoretical physics, it's time to come back to earth (or at least this version of it). To reflect, reassess and see if we have learned anything. In the course of this book we've covered numerous stages of the scientific process, from making an initial observation to focusing your research, working well with others, combating bias and hunting for proof. If it feels a tad overwhelming, then welcome to the club. Science is like that: you are constantly surrounded by reams of papers to read, endless sources of data to poke around in, processes to manage and people to please. To compound things even further, every field of research is dynamic, with new questions, data sets, technologies and theories constantly being added to the mix. It's close to impossible to keep up with your own work, let alone everyone else's.

As we have seen, the magnitude of science and its magnetic pull on those who love it can be both a blessing and a curse. It has inspired discoveries that have changed the world for the better, as well as leading researchers down rabbit holes and pushing them to seek evidence that will never be found, or embrace theories that cannot be proven. The genius idea and the crackpot belief can be two sides of the same coin.

It's my belief that we have so much to learn from people who flip that coin for a living, whichever side up it happens to land. It is not only scientists who have done amazing things that make good teachers, but those whose work has gone catastrophically wrong. Because science is a bit like life (after all, it's humans who are doing it), except optimized for what customer-service centres call 'quality and training purposes': people don't just do stuff, they also publish the results of their experiments, dig into what they did wrong and try and identify the most important things they have learned. (Try getting anyone to do that about their relationship history or tendency to have one drink too many.) It takes humility and bravery to dig further, which can often feel unnerving. But to make the world a better place, this is part of what's required.

From science we don't just get interesting stories and grand life lessons, but all the steps that led to that point. The scientific process demands a higher level of honesty and transparency about what decisions and mistakes were made, and which steps were followed towards the end goal. It's a good place to start if we want to understand what goes into spotting interesting problems, answering difficult questions and creating change. OK, so we have arrived at a solid place. We have done the legwork, found the sharpest tools in our drawer, and got the system on our side. But where does this leave us? What have we learned from how scientists work across a variety of different fields, and all different stages of the research process – from the glimmer of the first idea to the acceptance of a Nobel Prize?

You have to be passionate about your ideas – like Dr Katalin Karikó with mRNA vaccines – even when others are telling you to give up. The fact that (almost) nobody else believes in you or your idea is not necessarily a good enough reason to abandon it, if you really believe that you are onto something. But you also have to accept that, even with the best will in the world, a strong hypothesis and good evidence to back it up, your big idea might prove to be a dud. Like the astronomers who searched in vain for the planet Vulcan, the evidence you are seeking may never arrive, because it was never there to be found in the first place. Knowing when to quit can be as important as refusing to take an initial 'no' for an answer.

Whether you succeed or not in a particular pursuit, science also reminds us that you are unlikely to get anywhere alone. Remember that Karikó relied on her research partner, Dr Drew Weissman, to bring mRNA from theory to reality, and that the two brought contrasting skill sets to the table (one a biochemist and the other an immunologist). Similarly, the great physicist Lise Meitner recognized that she could not progress her work to demystify the structure of the atom without the expertise of an experimental chemist, Otto Hahn. It's no different in so many other fields of work: writers need editors, just as musicians need producers and architects need builders. Very little of what we do in life is a truly solo enterprise and, despite science's hermetic reputation, it exemplifies the importance of working together: pooling ideas, perspectives and expertise. Never be afraid to ask for help. It's what Einstein would have wanted.

Science also points to the need for a balance of certainty and flexibility in determining a course of action. While research demands a level of commitment to a particular subject or method, from all those you could possibly have chosen, there are also times that call for the plan to be thrown out of the window. Recall how Dr Charles Swanton started out by looking at how tumours resisted drug treatments, and then adapted his research, based on what he found, to look instead at their evolutionary patterns – a possibility that had only become clear once his team had started to get their hands dirty. Jobs, relationships, hobbies, and pretty much anything you care to name, answer to the same description. You can set a route, but must always be willing to alter it based on what you encounter along the way.

And, just as in science, you can expect the journey to any-thing meaningful to take a long time. Many of the examples we have explored show how big ideas in science are not the products of months or even years of work, but entire decades of research – with new steps often catalysed by the arrival of improved technology, policy, or ideas that come from other fields. It can take a long time for important scientific work to come together, and then for its full significance to gain traction and be recognized. Anyone impatient about achieving success in their life should reflect on how the Nobel Prize is often awarded to people deep into retirement: since 2007, four nonagenarians have become Laureates, the oldest being ninety-seven-year-old John Goodenough, recognized for his work in developing the lithium-ion battery.

Often the long-term nature of science means that work will be left incomplete. Richard Feynman died with unfinished work still chalked up on his blackboard, and Albert Einstein and Stephen Hawking were still searching for answers when they passed away, both at the age of seventy-six. Einstein never completed his search for a unified field theory that would connect our understanding of all the particles and forces in the universe, while Hawking was still poking around in the mysteries of black holes. We should all be as willing to make a lifetime pursuit of the things we care about most.

If switching course (and staying the course) is one imperative that science encourages, then another is to broaden your horizons. Sometimes it is not enough to think outside the box. You have to build a new one altogether, different from anything that has been tried before. Like the quantum theory of consciousness, which is not so much a solid hypothesis as a way of thinking that tries to encourage people to consider new possibilities about a subject we are only on the fringes of understanding. Sometimes the consensus thinking or method is insufficient for the task at hand. Science teaches us not to be afraid about straying beyond its boundaries, even when doing so will initially leave you with more questions than answers.

Trying to draw wider lessons from the study of science brings us to a whole hatful of paradoxical truths. To make progress, you have to think big and be bold, but also stress the fine print and think about everything that could go wrong and every potential flaw in your hypothesis. You have to pick a path from the plethora of options that are available to you, at the same time as being willing to veer off course if the evidence leads you there. You have to love consensus ideas and learn to work within them, even if your real ambition is to break new ground (whether as a Popper, trying to tear down the status quo, or as a Kuhn, only abandoning it when the anomalies become too great to bear). And you need to be enough of an individual to promote your own ideas and approaches, without being an individualist who is incapable of working with others or securing institutional support.

That might make working in science seem like a nightmare balancing act (which it can be), but it's also part of the attraction. Most scientists began as kids obsessed with how the world works, but they also got addicted to walking the wobbly bridge that takes you from observation to experiment, discovery and beyond. To the adrenaline rush of feeling the pieces fit together in your head, which is as good as scoring a goal in a football match (and which once happened to me while watching one – leading to some odd looks when I stood up and punched the air when nothing seemed to be happening). Almost all of the scientists I spoke to for this book, however advanced and successful in their career, reflected on this point. They do the work because it's not just what they know, but what they love. The thrill of getting a brainwave, spotting a telling detail or receiving an experimental result that fits like a glove never goes away, whether it's the first time or the hundredth. And the purpose of the work – from researching new treatments for deadly disease to transforming how we understand the universe and everything in it – speaks for itself. One of the comforts of scientific work, even when you are doing something seemingly mundane like crunching through a data set or running mice through a maze, is the thought that your work might make a contribution to something that changes or even saves someone's life.

Science shows us how to shoot for the moon (and frankly far beyond it), and that no one can stop you from pursuing an idea that you really believe in. Along the way, it provides a pretty handy guidebook to climbing mountains – metaphorical or literal – that feel insurmountable: choosing a path, focusing on your goal, adjusting as you go along, finding the fellow travellers you will need to succeed, and avoiding the many pitfalls of bias or belief that outrun evidence. I like to think of science as a combination of the arm around the shoulder and the kick in the pants that we all need to succeed. It encourages big ambitions without letting them fly away into the land of magical thinking. Do pursue your dreams, but don't expect they will come easily. Do challenge the consensus, but make sure you understand the rules you wish to break. Do believe in yourself, but don't discount your own biases and those of the people you are working with. As a scientist you have an incredible licence to explore ideas and try things out, but you will also be held accountable. You have to (or at least you should have to) show your workings, produce your evidence and let other people test drive your data. If you make big claims, expect people to try and poke holes straight through them. That combination of freedom and constraint is a pretty good way to think about most of the challenges we will encounter in life. No one is finally stopping you from having a go at what you really want to do, but equally nothing important ever comes easily. The more honest and accountable you are – both to yourself and other people – the better the results are likely to be. If you don't take the peer review seriously – like what the friends who have known you longest think about the person you think you have just fallen in love with – then don't blame anyone else if it all goes wrong.

Here's a final paradox about science that provides an important lesson in itself. It's a subject that is often associated with complexity, something that can be intimidating for people (including those who are professionally trained). Many of the scientific discoveries we have explored in this book are the product of work that was astonishingly clever and complicated, both in concept and practice. But here's the rub. Science is also about simplicity: distilling the contents of the world down into equations, principles and laws that allow us to consistently observe how things work. Science doesn't exist to make things complicated but to make an unimaginably large universe simple enough that we can begin to grasp what is really going on. That's why some of the most important, lasting ideas in science are straightforward enough that we teach them to primary schoolchildren: that two colliding objects exert an equal and opposite force on each other, that energy can neither be created nor destroyed, that nothing can move faster than the speed of light, and that we developed into *homo sapiens* through the process of evolution.

Most scientists are ultimately aiming for this kind of simplicity in their work: to take all the information and noise and try and boil it down to a solution, a theory, a product or a principle that is both fundamental and widely applicable. Simplicity is not the beginning but the end of the process – the reward for all your hard work, at which point everyone turns around and says: 'Why didn't I think of that?' – just like Niels Bohr when the principle of nuclear fusion was first put to him. Simplicity is something we should generally be aiming for in life too: a career we find fulfilling, relationships that bring us joy, interests that enrich us, the ability to help other people, and the material means to live without fear. Yet we have a tendency to over-complicate, favouring complex solutions or decision-making processes because the greater rigour involved gives us the comfort that they must be robust (even when the right answer was staring you in the face anyway).

We shouldn't be afraid to simplify, and we shouldn't shy away from what Leidy Klotz, Professor of Engineering and Architecture at the University of Virginia, calls subtraction – or the science of less.[1] His winningly simple idea is that our minds are geared to operate on the principle of more is better, rooted in our hunter-gatherer instincts to stockpile food as a means to survive. We are constantly looking to do more things, get more stuff, visit more places, acquire more knowledge and earn more money. We want the bigger house, the better job, the longer and more adventurous holiday. Our whole life is the pursuit of another scoop of ice cream.

Klotz came up with the idea of subverting all this when playing with his son. They were building a Lego model bridge, and as he moved to fetch a brick to even up one of the support towers, he saw his three-year-old solving the problem for him. He had simply removed a brick on the other side. Fascinated, Klotz Snr started digging deeper and soon confirmed his assumption that as humans we are biased towards addition over subtraction. Repeating the Lego experiment with adults, he and his co-researchers found that just 12 per cent opted to remove bricks from a structure. With a piece of music that needed improvement, people were three times more likely to add than remove notes. And, asked to zhuzh up a soup recipe, just two out of ninety participants used fewer ingredients than originally specified.[2]

As he has suggested, we can see examples of the benefits of subtraction – Marie Kondo-ing the mind – in all sorts of areas of life. Despite the belief that more is better, a team may function more efficiently and productively with fewer people in it. For the same reason, more companies are coming around to the idea that they might do better to have a working week of four days rather than five. What about films, where people have rarely sat through a three-hour epic without wondering whether it could have been shorter? Or, dare I say it, books: as every editor (hello, Emily!) will tell you, longer doesn't necessarily mean better.

Klotz doesn't just focus on benefits to the minutiae of life. He believes the principle of subtraction could make a major difference to how we think about our biggest problems, notably the climate crisis. As he suggests, our natural instincts are tending to lead us towards 'climate engineering' solutions, where we build and introduce things to the environment (such as dumping iron in the ocean to encourage carbon-sucking plant life to grow), but minimizing the role of subtraction – taking stuff away, such as the use of certain fertilizers in farming, buildings or structures that no longer serve their purpose (or are getting in the way of sustainable development), or even some of the masses of concrete that cover our cities. One of the great things about subtraction is that by its nature it requires you to consider the system first before taking away from it, not simply adding more layers without thinking about the indirect consequences and second-order effects. Subtraction in the context of climate is also a necessary challenge to individuals to think about what they can remove from their own carbon footprint, to question which aspects of the system aren't working, empowering them to understand their own role rather than feeling at the mercy of decisions made by corporations or believing that our only hope is for technological silver bullets. As the environmentalist Paul Hawken said to me: 'Climate change isn't a science problem but a human one.'[3]

Taking stuff away is a powerful idea that we could all apply to our lives in some way or form. It's also something that's fundamentally in tune with the way scientists work: hacking away at overgrown thickets of data, dispensing with anomalies and outliers, and trying to focus on the observations or experimental results that will take you somewhere – towards a small piece of what may one day be a big answer. But even that process is the product of a prior set of subtractions. Leidy Klotz further eluci-dated this when he told me: 'One thing I tell my students is that there's an infinite amount of valid questions, things that we don't know about the world, but there's not an infinite number of scientists and researchers. I hate this idea that "you can study anything and make a contribution to knowledge", [because] you can, but we have an obligation to figure out where we can make the biggest contribution to knowledge with the effort and the skills that we have.'[4]

We all face the same quandary, whether we are thinking about life, love, labour or leisure. Even our small corner of the world is pretty huge in terms of possibility. We can't have it all: like the PhD student trying to pick a research theme, we have to make decisions and – within reason – live by them. We have to work out what our ideal life looks like, and then get on with living it as best we can, trying not to worry about all the paths that have been spurned along the way. As the examples in this book have shown, finding the idea and picking a path is half the battle. Science gives us a perspective on how to do that, and it reminds us that even some of the smartest people you will meet have required bucketloads of patience, kindness, humility and honesty to make progress. And finally, it reassures us that we can be doing good work even when the results of it aren't imme-diately obvious. The prizes for all of your diligence and commitment may not be seen for some time. But as any scientist will tell you, doing interesting work, doing it properly and doing it to the best of your ability is a pretty great reward in itself and will always count for something.

So after thirty-one years in which I have mostly lived along-side science and been obsessed with it, I now look forward to the next thirty, my backpack full of the things I have learned, my notebook empty and a pen in my pocket. I'm not always sure what the next job or project is going to be, but I know that regardless of what I end up studying, where I live, or who is there to support me, I have everything I need to find my way. (Perhaps I'll have heard back on my grant application by then.)

Unlike the child who wrapped facts and piled books around her like a comfort blanket, as I have grown into an adult I have actually found out that I need to do things the other way round. To be able to prove myself wrong and adapt to the uncertainties of adult life have been the ultimate holy grail, and indeed a breakthrough for me, that will no doubt continue to unfold over the course of my life. Now as a researcher it is my duty to contribute, to devise experiments and test theories and help them stand up – to be part of the stability, not expect that science will provide it for me. To give back, even in small ways, is my definition of success and the real, raw condition of being human in the messy ecosystem in which we live.

Even though the role of science in my life has changed, its importance remains as strong as ever. I know that as I get older, develop some well-earned grey hairs and laughter lines, there will always be more questions to ask and more miniature obsessions to run after. I'll continue to get the squeaks of excitement that come from reading something interesting or encountering a question that can't easily be answered, and the motivating knots in my stomach when I can't figure a problem out. That is science in a nutshell: a magnet constantly drawing us towards new possibilities. It means there is always something more to question, to explore, to get excited about. Always a reason to have one foot out of the door and to be thinking about what comes next. And the knowledge that it won't be easy, that you will take wrong turns, be frustrated by your findings, and constantly get to the point of giving up hope? That, my friends, is all part of the parcel. All part of science's cocktail of hope and despair, excitement and frustration, success and failure. Sounds like fun to me. In fact, it sounds a little like life itself.

Notes

1. Observation

1 *Feynman Lectures on Computation: Anniversary Edition* (T. Hey, ed.), CRC Press, Florida, 2023, p. 360

2 Interview with Dr Ramin Hasani, 19 August 2021

3 Ibid.

4 D. Ackerman, '"Liquid" Machine-Learning System Adapts to Changing Conditions', *MIT News*, 28 January 2021, via www.news.mit.edu

5 L. Daston, 'On Scientific Observation', *Isis*, Vol. 99 (1), The University of Chicago Press, 2008, p. 97, via www.jstor.org

6 K. McQuater, 'David Spiegelhater: "Data does not tell you what to do"', Research Live, 16 March 2021, via www.research-live.com

7 K. Greff, S. van Steenkiste, J. Schmidhuber, 'On the Binding Problem in Artificial Neural Networks', ArXiv abs/2012.05208 (2020), p. 3

8 Ibid., p. 2

9 I. Goldstein, M. Goldstein, *The Experience of Science: An Interdisciplinary Approach*, Plenum Press, New York, 1984, p. 189

10 T. Greenhalgh, 'Miasmas, Mental Models and Preventative Public Health: Some Philosophical Reflections on Science in the COVID-19 Pandemic', *Interface Focus*, 11, 2021, p. 2, via www.royalsocietypublishing.org

11 L. Grossman, 'Dimension-Hop May Allow Neutrinos to Cheat Light Speed', *New Scientist*, 23 September 2011, via www.newscientist.com

12 National Human Genome Research Institute, 'The Cost of Sequencing a Human Genome', via www.genome.gov

13 MITCBMM, 'Liquid Neural Networks', via youtube.com: https://www.youtube.com/watch?v=IlliqYiRhMU

14 F. Balkwill, 'Cancers and the Tumour Microenvironment', Cambridge Society for the Application of Research lecture, 18 February 2019, via www.csar.org.uk

15 Interview with Professor Frances Balkwill, 16 September 2021

16 QMULOfficial, 'Building a 3D Model of Ovarian Cancer – Professors Fran Balkwill and Martin Knight', via youtube.com: https://www.youtube.com/watch?v=V61eD90JD48

17 J. Fricker, 'Tumour Microenvironment the New Battlespace in the War Against Cancer', *Cancerworld*, 83, Autumn 2018, p. 5, via www.archive.cancerworld.net

18 Ibid.

19 'Building a Human Tumour Microenvironment in the Lab', Cancer Research UK Barts Centre, 15 June 2021, via www.bartscancer.london

20 J. Willis and A. Todorov, 'First Impressions: Making Up Your Mind after a 100-Ms Exposure to a Face', *Psychological Science*, Vol. 17 (7), July 2006, via www.jstor.org

2. Hypothesis

1 Interview with Dr Katharina Schmack, 20 April 2022

2 Interview with Dr Chiara Marletto, 21 February 2022

3 Interview with Dr Charles Swanton, 9 December 2021

4 Ibid.

5 Ibid.

6 *Quanta Magazine*, 'The Theory that Could Rewrite the Laws of Physics', via youtube.com: https://www.youtube.com/watch?v=yboy6MRwmd4

7 C. Marletto, 'Life Without Design', *Aeon*, 16 July 2015, via www.aeon.co

8 C. Marletto, *The Science of Can and Can't: A Physicist's Journey Through the Land of Counterfactuals*, Allen Lane, London, 2021, p. xviii

9 Ibid.

10 Interview with Dr Chiara Marletto

11 K. Deisseroth, 'Optogenetics: Controlling the Brain with Light [Extended Version]', *Scientific American*, 20 October 2010, via www.scientificamerican.com

12 J. Colapinto, 'Lighting the Brain', *New Yorker*, 11 May 2015, via www.newyorker.com

13 Deisseroth, 'Optogenetics', op. cit.

3. Focus

1 C. McLarty, 'The Rising Sea: Grothendieck on Simplicity and Generality I', 24 May 2003

2 R. Galchen, 'The Mysterious Disappearance of a Revolutionary Mathematician', *New Yorker*, 9 May 2022, via www.newyorker.com

3 J. Cook, 'The Great Reformulation of Algebraic Geometry', John D. Cook Consulting, 25 September 2014, via www.johndcook.com

4 Enthought, 'UMAP: Uniform Manifold Approximation and Projection for Dimension Reduction | SciPy 2018', via youtube.com: https://www.youtube.com/watch?v=nq6iPZVUxZU

5 Interview with Dr Leland McInnes, 6 October 2021

6 Y. Hozumi, R. Wang et al., 'UMAP-assisted K-means clustering of large-scale SARS-Cov-2 mutation datasets', *Computers in Biology and Medicine*, Vol. 131, April 2021, via www.ncbi.nlm.nih.gov

7 A. Coenen, A. Pearce, 'Understanding UMAP': https://paircode.github.io/understanding-umap/

8 Interview with Dr Leland McInnes

9 M. Schwartz, 'The Importance of Stupidity in Scientific Research', *Journal of Cell Science*, 121, 1771, April 2008, via www.web.stanford.edu

10 E. Hoel, 'The Overfitted Brain: Dreams Evolved to Assist Generalization', *Patterns*, Vol. 2, Issue 5, 100244 (2021), pp. 3–7, via www.sciencedirect.com

11 Ibid., p. 7

12 M. Skokic, J. Collins et al., 'I Tried a Bunch of Things: the Dangers of Unexpected Overfitting in Classification', *Neuroscience and Biobehavioral Reviews*, Vol. 119, December 2020, pp. 456–7, via www.biorxiv.org

13 Ibid., pp. 9–10

14 Interview with Professor Rogier Kievit, 4 April 2022

15 R. Ratcliff and G. McKoon, 'The Diffusion Decision Model: Theory and Data for Two-Choice Decision Tasks', *Neural Computation*, Vol. 20 (4), April 2008, pp. 873–922, via www.ncbi.nlm.nih.gov

4. Interpret

1 S. Dresner, 'Polywater . . . the Water that Isn't', *Popular Science*, December 1969, p. 68

2 J. Stromberg, 'The Curious Case of Polywater', *Slate*, 7 November 2013, via www.slate.com

3 Stromberg, 'The Curious Case of Polywater', op. cit.

4 Denis L. Rousseau, 'Case Studies in Pathological Science', *American Scientist*, 80, no. 1, 1992, pp. 54–63, via www.jstor.org

5 Stromberg, 'The Curious Case of Polywater', op. cit.

6 Rousseau, 'Case Studies', op. cit., p. 57

7 Ibid., p. 54

8 I. Deary, 'An Intelligent Scotland: Professor Sir Godfrey Thomson and the Scottish Mental Surveys of 1932 and 1947', Joint

British Academy / British Psychological Society Lecture, 17 October 2012, via www.thebritishacademy.ac.uk

9 I. Deary, M. Whiteman and J. Starr, 'The Impact of Childhood Intelligence on Later Life: Following Up the Scottish Mental Surveys of 1932 and 1947', *Journal of Personality and Social Psychology*, 2004, Vol. 86, No. 1, pp. 130–47

10 A. Chuderski, 'The Broad Factor of Working Memory is Virtually Isomorphic to Fluid Intelligence Tested Under Time Pressure', *Personality and Individual Differences*, 85 (2015), pp. 98–104

11 Interview with Professor Rogier Kievit.

12 A. Luria, *Cognitive Development: Its Cultural and Social Foundations*, Harvard University Press, Cambridge, 1976, pp. 58–60

13 Ibid., p. 59

14 Interview with Professor Rogier Kievit.

15 'Obituary: Vera Rubin Died on December 25th', *The Economist*, 7 January 2017, via www.economist.com

16 R. Panek, *The 4% Universe: Dark Matter, Dark Energy, and the Race to Discover the Rest of Reality*, Oneworld, London, 2012, pp. 36–39

17 Ibid., p. 53

18 S. Perlmutter, 'Supernovae, Dark Energy, and the Accelerating Universe', *Physics Today*, Vol. 56 (4), 1 April 2003, via. pubs.aip.org

19 A. Riess, 'My Path to the Accelerating Universe', Nobel Lecture, Johns Hopkins University, 8 December 2011, via www.nobelprize.org

20 SLAC National Accelerator Laboratroy, 'Construction Begins on One of the World's Most Sensitive Dark Matter Experiments', phys.org, 7 May 2018, via www.phys.org

21 R. Panek, 'Dark Energy: The Biggest Mystery in the Universe', *Smithsonian Magazine*, April 2010, via www.smithsonianmag.org

22 Ibid.

23 Interview with Professor Sean Carroll, 9 February 2022

24 Riess, 'My Path to the Accelerating Universe', op. cit., pp. 13–15

25 'Vera Rubin', obituary in *The Times*, 31 December 2016, via www. thetimes.co.uk

26 Interview with Professor Sean Carroll

27 Riess, 'My Path to the Accelerating Universe', op. cit, p. 15

5. *Error, Failure and Troubleshooting*

1 Interview with Dr Chiara Marletto

2 M. Best, D. Neuhauser, 'Ignaz Semmelweis and the Birth of Infection Control', *BMJ Quality & Safety*, Vol. 13 (3), June 2004, pp. 233–4

3 Interview with Professor Frances Balkwill

4 T. Yu, 'How Scientists Drew Weissman and Katalin Karikó Developed the Revolutionary mRNA Technology Inside COVID Vaccines', *Bostonia*, 18 November 2021, via www.bu.edu

5 E. Dolgin, 'The Tangled History of mRNA Vaccines', *Nature*, 14 September 2021, via. www.nature.com

6 Dolgin, 'The tangled history', op. cit.

7 G. Kolata, 'Long Overlooked, Kati Karikó Helped Shield the World from the Coronavirus', *New York Times*, 8 April 2021, via www.nytimes.com

8 Ibid.

9 Yu, 'How Sciertists Drew Weissman and Katalin Karikó', op. cit.

10 P. Olliaro, 'What Does 95% COVID-19 Vaccine Efficacy Really Mean?', *The Lancet*, Vol. 21 (6), June 2021, p. 769, via www.thelancet.com

11 Kolata, 'Long Overlooked', op. cit.

12 The Nobel Assembly at Karolinska Institutet, 'The Nobel Assembly at the Karolinska Institutet has Today Decided to Award the 2023 Nobel Prize in Physiology or Medicine Jointly to Katalin Karikó and Drew Weissman', 2 October 2023, via www.nobelprize.org

13 Interview with Professor Massimo Pigliucci, 23 March 2022

14 C. Fishman, *The Big Thirst: The Secret Life and Turbulent Future of Water*, Free Press, New York, 2011, pp. 40, 325; L. Reynolds, 'The History of the Microwave Oven', *Microwave World*, vol. 10 (5), 1989, pp. 7–11

15 Y. Blanchard, G. Galati, P. van Genderen, 'The Cavity Magnetron: Not Just a British Invention', *IEEE Antennas and Propagation Magazine*, Vol. 55 (5), October 2013, p. 246, via www.ieeexplore.ieee.org

16 P. Allen, 'Scientists Paint Quantum Electronics with Beams of Light', *Uchicago News*, 9 October 2015, via news.uchicago.edu

6. Teamwork

1 K. Bird, M. Sherwin, *American Prometheus: The Triumph and Tragedy of J. Robert Oppenheimer*, Atlantic Books, London, 2009, p. 166

2 M. Harris, 'Overlooked for the Nobel: Lise Meitner', *Physics World*, 5 October 2020, via www.physicsworld.com

3 L. Meitner, 'Looking Back', *Bulletin of the Atomic Scientists*, Vol. 20 (9), November 1964, pp. 2–7, via www.aip.org

4 Ibid., p. 5

5 R. L. Sime, *Lise Meitner: A Life is Physics*, University of California Press, Los Angeles, 1997, p. 165

6 Ibid., p. 62

7 R. L. Sime, 'Lise Meitner and the Discovery of Nuclear Fission', *Scientific American*, Vol. 278 (1), January 1988, p. 84, via www.aip. org

8 R. L. Sime, *Lise Meitner: A Life in Physics* op. cit, p. 233

9 Ibid., p. 235

10 Ibid., p. 236

11 Ibid., p. 243

12 Bird, Sherwin, *American Prometheus*, op. cit., p. 166

13 H. Anderson, 'The Legacy of Fermi and Szilard', *The Bulletin of Atomic Scientists*, Vol. XXX, No. 7, p. 61.

14 Ibid., p. 62

15 'The Genius Behind the Bomb (1992)', via youtube.com: https://www.youtube.com/watch?v=OgT-Gw6Pjz4

16 'The Legacy of Fermi and Szilard', op. cit., p. 62; W. Lanouette, 'Einstein and Szilard in Princeton', The Lewis B. Cuyler Lecture, Historical Society of Princeton, 9 February 2011

17 D. Lewis, 'The World's First Nuclear Reactor Was Built in a Squash Court', *Smithsonian Magazine*, 25 November 2015, via www.smithsonianmag.com

18 C. Nelson, *The Age of Radiance: The Epic Rise and Dramatic Fall of the Atomic Era*, Scribner, New York, 2014, p. 130

19 Interview with Dr Leland McInnes

20 Ibid.

21 C. Hernandez, 'By Playing It Safe, I Became a Latino Scientist. But that Approach Held Me Back', *Science*, Vol. 374, Issue 6573, 10 December 2021, via. www.science.org

22 R. Prasad, 'Eight Ways the World Is Not Designed for Women', BBC News, 5 June 2019, via www.bbc.co.uk

23 M. Sjoding, T. Valley, T. Iwashyna, 'Racially Biased Oxygen Readings Could Be Putting Patients at Risk', Institute for Healthcare Policy & Innovation, University of Michigan, 16 December 2020, via www.ihpi.umich.edu

24 Interview with Dr Leland McInnes

25 Interview with Professor Rogier Kievit

26 The Royal Institution, 'How Does Science Work and Why Does It Matter? – with Jeremy Baumberg', via youtube.com: https://www.youtube.com/watch?v=foI5PP5EZvw

27 J. Baumberg, *The Secret Life of Science: How It Really Works and Why It Matters*, Princeton University Press, New, Jersey, 2018, pp. 13–14

28 Ibid., p. 14

29 'How Does Science Work?', op. cit.

7. Proof

1 J. C. Hafele, R. Keating, 'Around-the-World Atomic Clocks: Predicted Relatavistic Time Gains', *Science*, Vol. 177 (4044), July 1972, pp. 166–8

2 D. Castelvecchi, 'How Gravitational Waves Could Solve Some of the Universe's Deepest Mysteries', *Nature*, 11 April 2018, via nature.com

3 D. Overbye, 'Gravitational Waves Detected, Confirming Einstein's Theory', *New York Times*, 11 February 2016

4 Ibid.; 'Rainer Weiss: 50 Years of LIGO and Gravitational Waves', *Physics World*, 6 October 2022; K. Thorne and R. Weiss, 'A Brief History of LIGO', Caltech, 16 February 2016

5 K. Popper, 'Science as Falsification', via www.staff.washington.edu excerpted form T. Schick (ed.), *Readings in the Philosophy of Science*, Mayfield Publishing Company, Mountain View, 2002, pp. 9–13

6 T. Kuhn, *The Structure of Scientific Revolutions*, University of Chicago, Chicago, 1970 (2nd ed.), pp. 5–6

7 T. Levenson, *The Hunt for Vulcan: How Albert Einstein Destroyed a Planet, Discovered Relativity, and Deciphered the Universe*, Head of Zeus, London, 2016, pp. 73–4

8 Ibid., pp. 73–8

9 R. Harvey, 'Total Eclipse of the Sun', Bentley Historical Library, via https://bentley.umich.edu/

10 https://course-building.s3-us-west-2.amazonaws.com/Chemistry/transcripts/RichardFeynmanOnScientificMethod1964_transcript.txt; R. Feynman, 'Seeking New Laws', Messerger Lectures, Cornell University, 9 November 1964, via www.jamesclear.com

11 'Why Newtonian Gravity is Reliable in Large-Scale Cosmological Simulations', Monthly Notices of the Royal Astronomical Society, Oxford Academic (oup.com)

12 M. Gleiser, 'Unification' in *This Idea Must Die: Scientific Theories that are Blocking Progress* (J. Brockman, ed.), Harper Perennial, New York, 2015, p. 5

13 Ibid., p. 7

14 J. Travis, 'New Era in Digital Biology: AI Reveals Structures of Nearly All Known Proteins', 29 July 2022, science.org

15 Podcast: Episode 1, 'Predicting Protein Structure', Springer Nature Protocols and Methods Community

8. Bias

1 C. Sloan, 'Feathers for T. Rex? New Birdlike fossils are Missing Links in Dinosaur Evolution', *National Geographic*, Vol. 196 (5), November 1999, pp. 99–107, via www.archive.org; S. Austin, 'Archaeoraptor: Feathered Dinosaur from National Geographic Doesn't Fly', ICR, 1 March 2000, via www.icr.org

2 J. Pickrell, 'How Fake Fossils Pervert Paleontology [Extract]', *Scientific American*, 15 November 2014, via www.scientificamerican.com

3 E. Singer, 'How Dinosaurs Shrank and Became Birds', *Scientific American*, 12 June 2015, via www.scientificamerican.com

4 D. Schulze-Makuch, 'We Might Have Accidentally Killed the Only Life We Ever Found on Mars Nearly 50 Years Ago', *Big Think*, 27 June 2023

5 L. Samhita and H. Gross, 'The "Clever Hans Phenomenon" Revisited', *Communicative & Integrative Biology*, Vol. 6 (6), November–December 2013, via www.ncbi.nlm.nih.gov

6 R. Rosenthal and K. Fode, 'The Effect of Experimenter Bias on the Performance of the Albino Rat', *Behavioral Science*, Vol. 8 (3), 1963, pp. 183–9, via www.gwern.net

7 R. Rosenthal and L. Jacobson, 'Pygmalion in the Classroom', *Urban Review*, Vol. 3, September 1968, pp. 16–20, via www.sites.tufts.edu

8 R. Roper, 'Does Gender Bias Still Affect Women in Science?', *Microbiology and Molecular Biology Reviews*, Vol. 83 (3), 17 July 2019, via www.journals.asm.org

9 D. Kahneman, O. Sibony, C. Sunstein, *Noise: A Flaw in Human Judgment*, William Collins, London, 2021, pp. 249–50

10 Kahneman, Sibony, Sunstein, Noise, op. cit.

11 M. Aamodt, E. Kutcher et al., 'Do Structured Interviews Eliminate Bias? A Meta-analytic Comparison of Structured and Unstructured Interviews', presented at the annual meeting of the Society for Industrial-Organizational Psychology, May 2006, Dallas Texas, via researchgate.net

12 Kahneman, Sibony, Sunstein, *Noise*, op. cit., p. 305

13 Interview with Dr Alexander Wissner-Gross, 4 December 2021

14 TED, 'Alex Wissner-Gross: A New Equation for Intelligence', via youtube.com: https://www.youtube.com/watch?v=ue2ZEmTJ_Xo&t=519s

15 Interview with Dr Alexander Wissner-Gross

9. Imagination

1 N. Bostrom, 'Are You Living in a Computer Simulation?', *Philosophical Quarterly*, Vol. 53, No. 211, 2003, pp. 243–55, via www.ora.ox.ac.uk

2 D. Chalmers, *Reality+: Virtual Worlds and the Problems of Philosophy*, Penguin, London, 2023, pp. xiv–xvii; Descartes, *Meditations on First Philosophy with Selections from the Objections and Replies* (J. Cottingham ed.), Cambridge University Press, Cambridge, 2017 (2nd editon), p. 19

3 Chalmers, *Reality+*, op. cit., p. 4

4 S. Carroll, *Something Deeply Hidden: Quantum Worlds and the Emergence of Spacetime*, OneWorld Publications, New York, 2019, p. 1

5 Ibid., p. 18

6 F. Wilczek, 'Einstein's Parable of Quantum Insanity', *Quanta Magazine*, 23 September 2015, via www.scientificamerican.com

7 M. Brooks, 'Carlo Rovelli on the Bizarre World of Relational Quantum Mechanics', *New Scientist*, 10 October 2022, via www.newscientist.com

8 P. Byrne, 'The Many Worlds of Hugh Everett', *Scientific American*, 21 October 2008, via www.scientificamerican.com; S. Carroll, *Something Deeply Hidden*, op. cit.

9 'Everett's Letter to Bryce DeWitt of May 31, 1957', via www.pbs.org

10 Byrne, 'The Many Worlds', op. cit.

11 H. Everett (J. Barrett, P. Byrne eds), *The Everett Interpretation of Quantum Mechanics: Collected Works 1955–1980 with Commentary*, Princeton University Press, Princeton, 2012, p. 21

12 Ibid., pp 18–19

13 R. Galchen, 'Dream Machine: The Mind-Expanding World of Quantum Computing', *New Yorker*, 2 May 2011, via www.newyorker.com

14 D. Chalmers, *The Conscious Mind: In Search of a Fundamental Theory*, Oxford University Press, Oxford, 1996, pp. xi–xii

15 A. Seth, 'Consciousness: The Last 50 Years (and the Next)', *Brain and Neuroscience Advances*, Vol. 2, January–December 2018, via www.ncbi.nlm.nih.gov

16 A. Seth, 'The Hard Problem of Consciousness is Already Beginning to Dissolve', *New Scientist*, 1 September 2021, via www.newscientist.com

17 S. Hameroff, R. Penrose, 'Consciousness in the Universe: A review of the "Orch-OR" theory, *Physics of Life Reviews*, Vol. 11 (1), March 2014, pp. 39–78

18 M. Brooks, 'Roger Penrose: "Consciousness Must Be Beyond Computable Physics"', *New Scientist*, 14 November 2022, via www.newscientist.com

Conclusion

1 L. Klotz, *Subtract: The Untapped Science of Less*, St Martin's Press, London, 2021
2 Ibid., p. 27
3 Interview with Paul Hawken, 20 September 2022
4 Interview with Dr Leidy Klotz, 12 November 2021

Acknowledgements

This book has spanned so many different epochs of my life over the last ten years, and so I would like to thank all of my university friends and professors for their encouragement and support throughout my PhD at UCL and the University of Bristol.

I'd also like to thank all the incredible researchers that I have interviewed for this book over the last three years. Whether they are explicitly mentioned in the book or not, these wonderful people have sparked discussions on science and were the voices of cutting-edge research that linchpins this book: Ramin Hasani, Frances Balkwill, Charlie Swanton, Chiara Marletto, Katharina Schmack, Leland McInnes, Rogier Kievit, Sean Carroll, Michael Shermer, Massimo Pigliucci, Jeremy Baumberg, James Heath, James Maccabe, Rizwan Virk, Alexander Wissner-Gross, Russell Rockne, Janet Thornton, Leidy Klotz and Paul Hawken.

Thank you to the most incredible editors and publishing team who have helped me write and engineer this book from the splash of ideas into a piece I am proud of – Josh Davis, Emily Robertson, and my wonderful literary agent, Adam Gauntlett. To all of the Penguin team for your support, Natalie Wall, Mary Chamberlain and Olivia Mead.

A huge thanks goes to my family for their endless patience in watching me learn the hard lessons of being an adult. My dad and stepmum, Nay, for having me to live with them throughout the lockdowns and inspiring me each day to keep doing what I love (along with realizing that my dad and I share the same brain!). My mum for her kindness, and for always having faith in me regardless of what shape I have grown into – her acceptance is second to none – and Rob for always having my back.

My older sister, Lydia, my life guru – patient and unwavering, you have been my rock, always knowing how to put out emotional fires with wisdom and reason, and most importantly calling me out when I make a wrong move. I am a better person because of you. The family ecosystem wouldn't be complete without my brother-in-law, Roo, for his mindful and resourceful approach to life. And thank you to my younger siblings, Tiger, Lilly and Aggie – anything is possible. Dream big, guys.

Of course, thanks also go to my glorious and extraordinary friends who have been such a great support to me. Thank you for being there in my lonely writing times and reminding me of the joys we live for – my sunshine friends, Elodie Garceau, Alice Strachan, Claire Horsely, Alison Spilsbury and Clara Eckstein.

Thanks also to Wendy (yes, this is my toy poodle), a constant companion throughout the writing of this book – you have been with me through thick and thin. We can *finally* go to the park now.

And lastly, a huge thanks goes to Joe Simpson, for introducing me to the wonders of graphic novels and comic con, and for opening up my world so I could write again. Your support has been crucial to the completion of this book, and I'll always be grateful.

Index

Page references in *italics* indicate images.

humility 39–40, 52, 101, 209, 217
hygiene 101–2, 178, 180, 185–6
hypothesis 7, 8, 12, 16, 19, 20, 32, *34*,
    35–52, 56
  branched-evolution hypothesis
    40–44
  chaos and 36–7, 38, 42, 49, 51
  constructor theory 45–8
  courage and 52
  defined 35
  failure of 37–8, 39, 41, 43–4, 51–2
  focus and 39, 44, 56, 65–6
  humility and 39–40, 52
  hypothesis for how to make
    hypotheses 45–6
  inter-generational 49–51
  optogenetics and 49–51
  stories, power of and 36
  uncertainty in scientific process
    and 38–40

ignorance 64–5
imagination 48, *188*, 189–207
  artificial intelligence and 189
  consciousness and 203–6
  courage and 202
  habit and 201–2
  many-worlds interpretation
    (MWI) 197–201, 206
  multiverse theory 201
  Orchestrated objective reduction
    (Orch-OR) 204–6
  pathological science and 94
  predictive processing and 204
  quantum mechanics 192–207
  research, fundamental tenet of
    pioneering 206–7
  simulation theory 190–92

uncertainty and 191–2, 195–7
uncertainty principle 195–7
unfamiliar situations and 202
universal wave function 198, 200
'Viking' mission to Mars (1976)
  and 173
immune systems/immune reaction
  immunization 103–5, 106, 108
  tumour microenvironment
    (TME) and 25, 26, 27
intelligence
  abstract-reasoning tests and 82–6
  artificial general intelligence 15,
    183, 184
  artificial intelligence (AI) and 14,
    182–4
  bias and 183
  brain and 14, 82–6, 182–4
  crystallized intelligence 84
  fluid intelligence 84
  machine intelligence 17
  tumours and 40
interdisciplinary research 124
interpersonal level 121–2
interpretation 8, 13, *76*, 77–96, 124,
    134, 158, 171, 173
  abstract-reasoning tests 82–6
  ambitious hypothesis, misinter-
    preting evidence in favour of 81
  Big Crunch 90
  blessing, interpretation as a 96
  cosmological constant and 93
  dark matter/dark energy and
    88–94
  established views/establishment
    people and 82
  experiment context/design and
    82–7

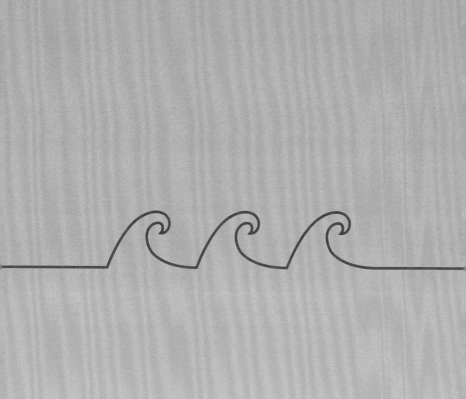